Worksheets

For Classroom or Lab Practice

Jeanette Shotwell

Central Texas College

Intermediate Algebra

FIFTH EDITION

Elayn Martin-Gay

PEARSON
Prentice Hall

Upper Saddle River, NJ 07458

Editorial Director, Mathematics: Christine Hoag
Editor-in-Chief: Paul Murphy
Sponsoring Editor: Mary Beckwith
Assistant Editor: Georgina Brown
Senior Managing Editor: Linda Mihatov Behrens
Associate Managing Editor: Bayani Mendoza de Leon
Project Manager, Production: Robert Merenoff
Art Director: Heather Scott
Supplement Cover Manager: Paul Gourhan
Supplement Cover Designer: Andrew Picone
Operations Specialist: Ilene Kahn
Senior Operations Supervisor: Diane Peirano
Mgr. Visual Research and Permissions: Karen Sanatar
Cover Image Credit: Scientific Illustrators / Laserwords

Pearson Prentice Hall
Pearson Education, Inc.
Upper Saddle River, NJ 07458

Printed in the United States of America
10 9 8 7 6 5 4 3 2 1

ISBN-13: 978-0-13-603105-5 Standalone
ISBN-10: 0-13-603105-6 Standalone
ISBN-13: 978-0-13-603107-9 Component
ISBN-10: 0-13-603107-2 Component

Pearson Education Ltd., London
Pearson Education Singapore, Pte. Ltd.
Pearson Education Canada, Inc.
Pearson Education—Japan
Pearson Education Australia PTY, Limited
Pearson Education North Asia, Ltd., Hong Kong
Pearson Educación de Mexico, S.A. de C.V.
Pearson Education Malaysia, Pte. Ltd.
Pearson Education Upper Saddle River, New Jersey

Worksheets

For Classroom or Lab Practice

Intermediate Algebra, Fifth Edition

Table of Contents

Name:

Date:

Instructor:

Section:

Chapter 1 Real Numbers and Algebraic Expressions
Section 1.2 Algebraic Expressions and Sets of Numbers

Learning Objectives

1. Identify and evaluate algebraic expressions.
2. Identify natural numbers, whole numbers, integers, and rational and irrational numbers.
3. Find the absolute value of a number.
4. Find the opposite of a number.
5. Write phrases as algebraic expressions.

Vocabulary

Use the choices to complete each statement.

Absolute value	**Expression**	**Integers**
Irrational number	**Natural numbers**	**Rational numbers**
Whole Numbers	**Variables**	

1. The set of ______________________ is {1, 2, 3, 4,…..}.
2. The number π is a(n) ______________________.
3. Any number that can be written as a fraction is a ______________________.
4. The ______________________ is the distance a number is from 0 on the number line.
5. The set of ________________ contains the positive and negative whole numbers.
6. ____________________ are letters that represent numbers.
7. The set of ______________________ is {0, 1, 2, 3, …..}
8. A collection of numbers, variables, and operations is known as a(n) ________________.

Objective 1

Evaluate each expression when $x = 2$ and $y = -3$.

9. xy

9. ______________________

10. $4x - 5y$

10. ______________________

11. $x^2 - 3xy + 9y^2$

11. ______________________

Name: Date:

Instructor: Section:

Objective 2

List the elements in each set.

12. $\{x|x$ is an even natural number less than $15\}$

12. ____________

13. $\{x|x$ is a whole number less than $1\}$

13. ____________

Place $\in$ or $\notin$ in the space provided to make each statement true.

14. -25 ____ $\{x|x$ is a whole number$\}$

14. ____________

15. 12 _____ $\{1, 3, 5, \ldots\}$

15. ____________

Objective 3

Find the absolute value.

16. $\left|-\frac{3}{5}\right|$

16. ____________

17. $-|2|$

17. ____________

18. $-|-16.4|$

18. ____________

Name: **Date:**

Instructor: **Section:**

Objective 4

Find the opposite of each.

19. 5.67

19. ____________________

20. $-\frac{4}{7}$

20. ____________________

21. -67

21. ____________________

Objective 5

Write each phrase as an algebraic expression. Use the variable x to represent each unknown number.

22. Five more than two-thirds a number.

22. ____________________

23. Twelve, minus three times a number.

23. ____________________

24. Twice the sum of eight and a number.

24. ____________________

Concept Extension

True or False.

25. The sum of two natural numbers will result in a natural number.

25. ____________________

26. The quotient of two integers will result in whole number.

26. ____________________

Name: **Date:**

Instructor: **Section:**

Section 1.3 Operations on Real Numbers

Learning Objectives

1. Add and subtract real numbers.
2. Multiply and divide real numbers.
3. Evaluate expressions containing exponents.
4. Find the roots of numbers.
5. Use the order of operations.
6. Evaluate algebraic expressions.

Objective 1

Add or subtract as indicated.

1. $24-(-16)$

1. ______________

2. $-18+(-19)$

2. ______________

3. Subtract $4/5$ from $1/3$.

3. ______________

Objective 2

Multiply or divide as indicated.

4. $(-2.4)(1.1)$

4. ______________

5. $3/5 \div -2/7$

5. ______________

6. $-3(-4)(-8)$

6. ______________

Name:

Instructor:

Date:

Section:

Objective 3

Evaluate each expression.

7. -7^2

7. ____________________

8. $\left(\frac{2}{5}\right)^3$

8. ____________________

9. $(-0.1)^5$

9. ____________________

Objective 4

Find the following root.

10. $\sqrt{36}$

10. ____________________

11. $\sqrt[5]{-32}$

11. ____________________

12. $\sqrt{-100}$

12. ____________________

Objective 5

Simplify each expression.

13. $(-3)^2 - 5(-4) + \sqrt{81}$

13. ____________________

Name:

Date:

Instructor:

Section:

14. $\dfrac{(3-\sqrt{9})-(-5-1.3)}{-3}$

14. ____________________

15. $-4\{3^2-2^3[5-2(-6+3)]\}$

15. ____________________

16. $\dfrac{5+2|8-13|}{6+2(-2)^2}$

16. ____________________

Objective 6

Evaluate each expression when $x = -5$ and $y = 4$.

17. $3x-4y$

17. ____________________

18. $\dfrac{8+2\sqrt{y}}{(x)^2}$

18. ____________________

19. $|5x+6y|-\sqrt{x^3+5y-1}$

19. ____________________

Name:

Date:

Instructor:

Section:

Concept Extension

20. Explain the difference between -5^2 and $(-5)^2$.

20. ____________________

21. What number needs to be added to the sum of $^3/_8$ and $^7/_{12}$ to make the result 1?

21. ____________________

Name: **Date:**

Instructor: **Section:**

Section 1.4 Properties of Real Numbers

Learning Objectives

1. Use the operation and order symbols to write mathematical sentences.
2. Identify identity numbers and inverses.
3. Identify and use the commutative, associative, and distributive properties.
4. Write algebraic expressions.
5. Simplify algebraic expressions.

Objective 1

Write each sentence using mathematical symbols.

1. Twice x, plus 5, is the same as -14.

1. ____________________

2. Three times the difference of *n* and five is less than fifteen.

2. ____________________

3. The quotient of *r* and negative nine is not equal to thirty.

3. ____________________

4. One-half of *t* is equal to the product of two-thirds and eighteen.

4. ____________________

Objective 2

5. Find the opposite and the reciprocal of 0.

5. ____________________

6. Find the opposite and the reciprocal of $7/9$.

6. ____________________

Name:

Instructor:

Date:

Section:

7. Find the opposite and the reciprocal of - 4.

7. ____________________

8. Find the opposite and the reciprocal of $-\frac{11}{18}$

8. ____________________

Objective 3

Use the commutative property to write an equivalent expression.

9. $x \bullet 3$

9. ____________________

10. $7x + 8y$

10. ____________________

Use the associative property to write an equivalent expression.

11. $8x + (7y + 3z)$

11. ____________________

12. $24(xy)$

12. ____________________

Use the distributive property to find the product.

13. $-4(x - y)$

13. ____________________

14. $0.25(t + 4s)$

14. ____________________

Name: | Date:
Instructor: | Section:

Objective 4

Write each of the following as an algebraic expression.

15. Two numbers have a sum of 112. If one number is x, represent the other number as an expression in x.

15. ______________________

16. The perimeter of a rectangle with width of w and length of 8.

16. ______________________

17. The cost of n movies if each movie is \$19.95.

17. ______________________

Concept Extension

Simplify each expression.

18. $\frac{1}{2}(10x-2)-\frac{1}{6}(60x-5y)$

18. ______________________

19. Show that subtraction is not commutative.

19. ______________________

20. Show that division is not commutative.

20. ______________________

Name: **Date:**

Instructor: **Section:**

Chapter 1 Vocabulary

Vocabulary Word	Definition	Example
Variable	Letter that represents a number.	$x,\ r,\ s$
Algebraic expression	Collection of numbers, operations, and variables.	$3x-7y^2$
Natural numbers	Counting numbers	{1, 2, 3,….}
Whole numbers	0 and the natural numbers	{0, 1, 2, 3,….}
Integers	Positive and negative numbers	{…-3. -2, -1, 0, 1, 2, 3, ….}
Rational numbers	Numbers that can be written as fractions.	$-4,\ \frac{5}{6},\ -0.25$
Irrational numbers	Numbers that cannot be written as fractions.	$\sqrt{3},$
Real numbers	Any number that can be found on the number line.	$-6,\ \frac{3}{4},$
Square root	$\sqrt{a}=b$ if $b^2=a$ b is positive.	$\sqrt{36}=6$
Identity	a + 0 = a	6 + 0 = 6
Inverse	$a+(-a)=0$ $a\left(\frac{1}{a}\right)=1$	$4+(-4)=0$ $4\left(\frac{1}{4}\right)=1$
Commutative property	$a+b=b+a$ $ab=ba$	$5+6=6+5$ $4(2)=2(4)$
Associative property	$a+(b+c)=(a+b)+c$ $a(bc)=(ab)c$	$1+(5+3)=(1+5)+3$ $4(2\bullet 3)=(4\bullet 2)3$
Distributive property	$a(b\pm c)=ab\pm ac$	$4(2+5)=8+20$

Name: **Date:**

Instructor: **Section:**

Practice Test A

Simplify.

1. $8+16\div(-4)-5^2$

1. ____________________

2. $\dfrac{13(8-2)-10(-8+2)}{2^3}$

2. ____________________

3. $20-\left\{15(3-4)+\sqrt{25}\right\}$

3, ____________________

4. $|8-10|^3-(-12+16)^2$

4. ____________________

5. Evaluate $\dfrac{x^2-5x}{8y}$ when $x=3$ and $y=-2$.

5. ____________________

Write each statement using mathematical symbols.

6. Ten times the sum of q and negative eight.

6. ____________________

7. The difference of seven and the square root of n is greater than or equal to nine.

7. ____________________

Name: **Date:**

Instructor: **Section:**

8. The absolute value of the product of three and t is not equal to eight more than one-half of t.

8. ______________________

Name the property illustrated.

9. $\frac{1}{2}+\left(-\frac{1}{2}\right)=0$

9. ______________________

10. $(3z+4y)+8x=3z+(4y+8x)$

10. ______________________

11. $8(3+x)=8(x+3)$

11. ______________________

12. Write an expression for the area of a triangle if the base is b, and height is twenty.

12. ______________________

Simplify each expression.

13. $\frac{3}{4}x-\frac{7}{8}+\frac{1}{6}x-\frac{2}{3}$

13. ______________________

14. $5c+10-4(c+8)$

14. ______________________

Name: Date:

Instructor: Section:

15. $-(17.4-7.2x)+(16.3-8.4x)$

15. ____________________

Insert <, >, or =.

16. -9 _____ -19

16. ____________________

17. -17 ______ $-|17|$

17. ____________________

List the elements in each set.

18. {x | x is an even whole number less than twelve}

18. ____________________

Find the absolute value.

19. $-|-25|$

19. ____________________

Find the following root.

20. $\sqrt{-81}$

20. ____________________

Name: Date:

Instructor: Section:

Practice Test B

1. Find the multiplicative inverse of $-\frac{8}{9}$.

a. $\frac{9}{8}$ b. $\frac{8}{9}$

c. $-\frac{9}{8}$ d. $-\frac{8}{9}$

1. ____________

2. Find the opposite of – 15.

a. 15 b. $\frac{1}{15}$

c. $-\frac{1}{5}$ d. – 15

2. ____________

3. True or false : π is a real number.

a. True b. False

3. ____________

4. $\dfrac{17+2^3}{\frac{1}{2}(10)-\sqrt{25}}$

a. $\frac{5}{2}$ b. undefined

c. 5 d. 1

4. ____________

5. $\left(\dfrac{2}{3}+\dfrac{1}{4}\right)^2-\dfrac{7}{12}$

a. $-\dfrac{49}{144}$ b. $\dfrac{25}{12}$

c. $\dfrac{37}{144}$ d. $\dfrac{205}{144}$

5. ____________

Name: **Date:**

Instructor: **Section:**

6. $\left|8^2-4^2\right|-(7-3)^2$

6. ____________________

a. 32

b. – 132

c. 112

d. 16

7. $\dfrac{\sqrt{10^2-3(25)}}{\sqrt{169}}$

7. ____________________

a. $25/13$

b. $5/13$

c. $13/5$

d. $-5/13$

8. $(4.8x-1.7)-(10.2x-2.9)$

8. ____________________

a. 15x + 1.2

b. 15x + 4.6

c. -5.4x + 1.2

d. -5.4x + 4.6

9. $-8(7x-3)-\left|18x-30x\right|$

9. ____________________

a. -44x + 24

b. -68x + 24

c. -44x - 24

d. -68x + 3

Name the property illustrated.

10. $(2+y)+4=2+(y+4)$

10. ____________________

a. Commutative property of addition

b. Associative property of addition

c. Distributive property

d. Commutative property of multiplication

Name: **Date:**

Instructor: **Section:**

11. $\left(\frac{1}{2}\right)2 = 1$

11. ____________

a. Distributive property b. Multiplicative inverse property.

c. Multiplication Identity property d. Associative property of multiplication

12. $24x - 48 = 24(x - 2)$

12. ____________

a. Distributive property b. Associative property of addition

c. Commutative property of addition d. Associative property of multiplication

Write each expression using mathematical symbols.

13. The sum of nine and p is less than two.

13. ____________

a. $9 + p - 2$ b. $9 + p < 2$

c. $9 + p \leq 2$ d. $9 < p + 2$

14. Twelve subtracted from $\frac{1}{2}$ of x is not equal to fourteen.

14. ____________

a. $\frac{1}{2}x - 6 \neq 14$ b. $6 - \frac{1}{2}x \neq 7$

c. $\frac{1}{2}x - 12 = 14$ d. $\frac{1}{2}x - 12 \neq 14$

15. The absolute value of the sum of t and four is equal to five.

15. ____________

a. $|t + 4| = 5$ b. $|t| + 4 = 5$

c. $t + 4 = |5|$ d. $|t + 4| \neq 5$

Name:

Instructor:

Date:

Section:

Evaluate each expression.

16. -8^2

16. ____________________

a. 16 b. 64

c. –16 d. – 64

17. $\frac{-17}{0}$

17. ____________________

a. 0 b. ∅

c. 17 d. – 17

18. $\sqrt[5]{-32}$

18. ____________________

a. –2 b. 2

c. not a real number d. 32

19. The formula to convert Celsius to Fahrenheit is $f = 1.8c + 32$. What would $-8°$ C be in Fahrenheit?

19. ____________________

a. 46.4° b. 33.8°

c. 17.6° d. 208°

20. Write an expression for the total money of *x* quarters and *y* dimes. (in dollars)

20. ____________________

a. 25x + 10y b. 0.25x + 0.10y

c. 0.35xy d. 0.10x + 0.25y

Name: Date:

Instructor: Section:

Chapter 2 Equations, Inequalities, and Problem Solving
Section 2.1 Linear Equations in One Variable

Learning Objectives
1. Solve linear equations using properties of equality.
2. Solve linear equations that can be simplified by combining like terms.
3. Solve linear equations containing fractions or decimals.
4. Recognize identities and equations with no solutions.

Vocabulary

Use the choices to complete each statement.

Addition	**Equivalent**	**Like**
Multiplication	**Solution**	**Value**

1. A(n) ________________ of an equation, is a value for the variable that will make the equation true.
2. ________________ equations are equations that have the same solution.
3. By the ________________ property of equality, $8x = 64$ and $\frac{8x}{8} = \frac{64}{8}$ are equivalent equations.
4. By the ____________ property of equality, $x + 4 = 9$ and $x + 4 - 4 = 9 - 4$ are equivalent equations.

Objective 1

Solve each equation.

5. $-8x = 24$

5. ____________________

6. $4.1 - 7z = 3.6$

6. ____________________

7. $3x + 7 = 2x - 14$

7. ____________________

8. $z + 14.7 = 18.2$

8. ____________________

Name: Date:

Instructor: Section:

Objective 2

Solve each equation.

9. $6(x-2)=3x+10$

9. ____________________

10. $-3(x-9)=5x-3(x+2)$

10. ____________________

11. $17-2x+18+3x=-4x-15+10x$

11. ____________________

12. $-10(y-2)=-15(y-2)$

12. ____________________

Objective 3

Solve each equation.

13. $\frac{1}{2}(x-4)=\frac{3}{4}x+1$

13. ____________________

14. $\frac{x+3}{4}+\frac{x-2}{3}=\frac{1}{6}$

14. ____________________

Name:

Instructor:

Date:

Section:

15. $0.5x - 0.4 = 0.2(x-1)$

15. ____________________

16. $0.25x + 0.80 = -0.4(x-1)$

16. ____________________

17. $\frac{m-4}{3} - \frac{3m-1}{5} = 1$

17. ____________________

Objective 4

Solve each equation.

18. $3(x+1)+5 = 3x+2$

18. ____________________

19. $4(x+2) = 3(4x+2) - 2(4x-1)$

19. ____________________

20. $4(x+1) - x = 8(x-1) - 5x$

20. ____________________

Name: **Date:**

Instructor: **Section:**

21. $3(2x+3)-(-x-2)=4(3x+5)-(x+9)$

21. ____________________

Concept Extensions

Solve each equation.

22. $7.342x=-19.0892$

22. ____________________

23. $\frac{5}{12}x+3=\frac{7}{12}x-11$

23. ____________________

Name: **Date:**

Instructor: **Section:**

Section 2.2 An Introduction to Problem Solving

Learning Objectives

1. Write algebraic expressions that can be simplified.
2. Apply the steps for problem solving.

Objective 1

Write the following as algebraic expressions. Then simplify.

1. The area of a rectangle with a length of x and width of $x-2$.

1. ____________________

2. The sum of four consecutive even integers.

2. ____________________

3. The total amount of money (in cents) in x nickels, $x+3$ dimes, and $2x$ quarters.

3. ____________________

4. The area of a triangle with base of $2x$ meters and height of $x+4$ meters.

4. ____________________

5. The circumference of a circle whose diameter is $x+4$.

5. ____________________

Objective 2

Solve.

6. A second number is five times a first number. A third number is 100 more than the first number. If the sum of the three numbers is 415, find the numbers.

6. ____________________

Name: **Date:**

Instructor: **Section:**

7. The sum of three consecutive integers is 75. Find the integers.

7. ____________________

8. One angle is $20°$ more than four times its supplement. Find the angles.

8. ____________________

9. In a recent survey, 40% of a developmental math class were females. If there were a total of 240 males in all the developmental classes, how many were females?

9. ____________________

10. Find the measures of the angles of a triangle if the measure of one angle is ten less than the second angle and the third angle is three times the first angle.

10. ____________________

11. A computer was purchased for $594.12 including tax. If the tax rate was 6% what was the price of the computer.

11. ____________________

12. Two frames are needed with the same perimeter: one frame in the shape of a square and one in the shape of an equilateral triangle. Each side of the triangle is 6 centimeters longer than each side of the square. Find the dimensions of each frame.

12. ____________________

Name: **Date:**

Instructor: **Section:**

Concept Extensions

Solve.

13. How many cubic centimeters of a 30% solution should be mixed with an 80% solution to get 50 cubic centimeters of a 60% solution?

13. ____________________

Name: **Date:**

Instructor: **Section:**

Section 2.3 Formulas and Problem Solving

Learning Objectives

1. Solve a formula for a specified variable.
2. Use formulas to solve problems.

Objective 1

Solve each equation for the specified variable.

1. $y = mx + b$; for x.

1. ____________________

2. $F = ma$; for m.

2. ____________________

3. $F = \frac{9}{5}C + 32$; for C.

3. ____________________

4. $4x + 6y = 12$; for x

4. ____________________

5. $S = \frac{n}{2}(a + L)$; for L.

5. ____________________

6. $y = ax^2 + bx + c$; for b.

6. ____________________

Name: **Date:**

Instructor: **Section:**

7. $S = B + \frac{1}{2}pl$; for *L*.

7. ______________________

Objective 2

Solve.

8. If it takes 3 hours to travel to Dallas, TX a total trip of 192 miles. How fast was the driver going?

8. ______________________

9. The surface area of a tomato sauce can is 52 square inches. If the company wants to make a can that has a diameter of 3 inches. How tall will the can be? Round to the nearest integer.

9. ______________________

10. If the area of a triangular kite is 18 square feet and its base is 4 feet, find the height of the kite.

10. ______________________

11. A banker invested $3000 in an account that earns 6% interest that is compounded quarterly. How much interest will be earned in 3 years?

11. ______________________

Name: **Date:**

Instructor: **Section:**

12. The low temperature in Las Vegas, NV was 14° F. What is the temperature in Celsius?

12. ____________________

13. The perimeter of a parking lot is 200 meters. If the length is 10 meters more than the width, find the dimensions of the parking lot.

13. ____________________

14. Given the perimeter of an isosceles right triangle is 127.5 inches and the hypotenuse is 49.5 inches. What is the measures of the legs?

14. ____________________

Concept Extensions

Solve.

15. The formula $E = mc^2$, where E = energy, m = mass , and c = speed of light. The speed of light is 1.8×10^5 meters per second squared. If the energy produced by an apple is $2.4 \times 10^{12}\ kg{\bullet}m^2/s^2$, what is the mass of the apple.

15. ____________________

Name: **Date:**

Instructor: **Section:**

Section 2.4 Linear Inequalities and Problem Solving

Learning Objectives

1. Use interval notation.
2. Solve linear inequalities using the addition property of inequality.
3. Solve linear inequalities using the multiplication and the addition properties of inequality.
4. Solve problems that can be modeled by linear inequalities.

Objective 1

Graph each set on a number line and then write in interval notation.

1. $\{x \mid x > 3\}$

1. ____________________

2. $\{x \mid x \leq -4\}$

2. ____________________

3. $\{x \mid -2 < x \leq 1.5\}$

3. ____________________

Objective 2

Solve. Write your answer in interval notation.

4. $x - 8 \geq 12$

4. ____________________

5. $x + 7 < 1$

5. ____________________

6. $x - 3 > -8$

6. ____________________

Name: Date:

Instructor: Section:

7. $4 + x \le 13$

7. ____________________

Objective 3

Solve. Write your answer in interval notation.

8. $2x > 42$

8. ____________________

9. $-3x - 7 \le 2$

9. ____________________

10. $\frac{3}{4}x - 1 > 7$

10. ____________________

11. $3(x-5) < 2(2x-1)$

11. ____________________

12. $7x + 2 \ge 3x - 1$

12. ____________________

13. $\frac{5x+1}{7} - \frac{2x-6}{4} \ge -4$

13. ____________________

Name:

Instructor:

Date:

Section:

Objective 4

Solve. Write your answer with an inequality.

14. A bride has a budget of \$1500 for her reception dinner. It costs \$28 per person for the meal she wants to serve. Use an inequality to find the number of guests the bride can invite to the reception.

14. ______________________

15. A small plane's maximum take-off weight is 2000 pounds or less. Six passengers weigh an average of 160 pounds each. Use an inequality to find the luggage and cargo weights the plane can carry.

15. ______________________

16. A rental car company offers two deals for their weekend special. (Weekend means 2 day rental)

    Plan A: \$32 per day with unlimited miles.
    Plan B: \$26 per day plus \$0.10 per mile.

    Use an inequality to find the number of daily miles for which Plan B is more economical than Plan A.

16. ______________________

Concept Extension

17. Why is it wrong for an interval notation to be written as $(-4, -\infty)$?

17. ______________________

18. Write an inequality that includes both multiplication and addition whose solution is $\{x \mid x \geq -3\}$.

18. ______________________

Name: Date:

Instructor: Section:

Section 2.5 Compound Inequalities

Learning Objectives

1. Find the intersection of two sets.
2. Solve compound inequalities containing and.
3. Find the union of two sets.
4. Solve compound inequalities containing or.

Vocabulary

Use the choices to complete each statement.

Compound	**Intersection**	**Union**
$\cap$	$\cup$	

1. An inequality that contains two inequalities is called a(n) ________________ inequality.
2. The word and means ____________ and uses the symbol ____________.
3. The word or means ____________ and uses the symbol ____________.

Objective 1

If $A = \{x \mid x$ is an odd integer between 0 and 30$\}$, $B = \{x \mid x$ is an even number$\}$, $C = \{2, 5, 7, 13, 18, 22, 27\}$, and $D = \{1, 6, 12, 15, 19, 23, 26, 29\}$, list the elements of each set.

4. $C \cap B$

4. ____________________

5. $D \cap A$

5. ____________________

6. $C \cap A$

6. ____________________

7. $C \cap D$

7. ____________________

Name:

Instructor:

Date:

Section:

Objective 2

Solve each compound inequality. Write your answer in interval notation.

8. $x \geq 3$ and $x < 10$

8. ______________________

9. $x \leq 8$ and $x \geq -4$

9. ______________________

10. $x+1 \geq 7$ and $3x-1 \geq 5$

10. ______________________

11. $-6x \leq -18$ and $x-20 \leq 10$

11. ______________________

12. $-4 \leq x+3 < 11$

12. ______________________

13. $1 \leq \frac{2}{3}x+3 \leq 4$

13. ______________________

Name: **Date:**

Instructor: **Section:**

Objective 3

If $A=\{x \mid x \text{ is an odd integer}\}$, $B=\{x \mid x \text{ is an even number}\}$, $C=\{2,9,12,19\}$, and $D=\{3,8,13,18\}$, list the elements of each set.

14. $C \cup D$

15. $A \cup C$

16. $B \cup D$

14. ____________________

15. ____________________

16. ____________________

Objective 4

Solve each compound inequality. Write your answer in interval notation.

17. $x<2$ or $x<-3$

18. $x \geq 5$ or $x \geq -5$

19. $2x>12$ or $x+7>3$

20. $4(x-1)<9$ or $5(x+3) \leq 12$

17. ____________________

18. ____________________

19. ____________________

20. ____________________

Name: **Date:**

Instructor: **Section:**

21. $x < 1$ or $x > 5$

21. ____________________

22. $x - 7 > -4$ or $-2x - 14 \geq 20$

22. ____________________

Concept Extension

Solve each compound inequality. Write your answer in interval notation.

23. $1 - 3x < 4x + 5 \leq x + 15$

23. ____________________

24. $4(2 + 3x) \leq 2(x + 2) \leq -3(x - 2)$

24. ____________________

Name: **Date:**

Instructor: **Section:**

Section 2.6 Absolute Value Equations

Learning Objectives

1. Solve absolute value equations.

Objective 1

Solve each absolute value equation.

1. $|x| = 8$

1. ____________________

2. $|x+2| = 10$

2. ____________________

3. $|2x| - 4 = 16$

3. ____________________

4. $|z| = -2$

4. ____________________

5. $|2x-1| = 5$

5. ____________________

Name:

Instructor:

Date:

Section:

6. $|x-4|+3=10$

6. ____________________

7. $\left|\frac{x}{2}-3\right|=1$

7. ____________________

8. $|1+2x|=|3-x|$

8. ____________________

9. $|x|=|x-4|$

9. ____________________

10. $|2y-3|=|9-4y|$

10. ____________________

Name:

Instructor:

Date:

Section:

11. $2-|3n|=8$

11. ______________________

12. $\left|\frac{x+1}{8}\right|=\left|\frac{x-2}{4}\right|$

12. ______________________

Concept Extensions

13. Write an absolute value equation representing all numbers x whose distances from – 5 is 4 units.

13. ______________________

14. For what value(s) of c will an absolute value equation of the form $|ax-b|=c$

A. one solution?
B. no solution?
C. two solutions?

14a. ______________________

14b. ______________________

14c. ______________________

Name:

Instructor:

Date:

Section:

Section 2.7 Absolute Value Inequalities

Learning Objectives

1. Solve absolute value inequalities of the form $|x| < a$.
2. Solve absolute value inequalities of the form $|x| > a$.

Objective 1

Solve each inequality. Write your answer in interval notation.

1. $|x| < 5$

1. ____________________

2. $|3x| \leq 18$

2. ____________________

3. $|x-3| < 6$

3. ____________________

4. $7 - |x+1| > 4$

4. ____________________

5. $\left|\frac{x+1}{3}\right| \leq 5$

5. ____________________

Name: Date:

Instructor: Section:

6. $|x-5|<-4$

6. ____________________

7. $|3x-1|<-5$

7. ____________________

8. $|2x-1|+4\leq 9$

8. ____________________

9. $-15+|2x-7|\leq -6$

9. ____________________

Objective 2

Solve each inequality. Write your answer in interval notation.

10. $|x|>-9$

10. ____________________

11. $|x-1|\geq 5$

11. ____________________

12. $|1-2x|>6$

12. ____________________

Name: **Date:**

Instructor: **Section:**

13. $\left|\frac{x-3}{4}\right| > 6$

13. ____________________

14. $\left|\frac{x+6}{3}\right| > 2$

14. ____________________

15. $12 - |x-3| \le 7$

15. ____________________

16. $|2x-3| \ge 7$

16. ____________________

Name: Date:

Instructor: Section:

Chapter 2 Vocabulary

Vocabulary Word	Definition	Example
Equation	A statement that two expressions are equal	$3x-1=6$
Solution	A value of the variable that makes the equation true.	2 is a solution of $x+4=6$
Equivalent equation	Equations that have the same solution.	$x+4=8$ and $2x=8$ are equivalent.
Addition property of equality	The process of adding/subtracting a constant from each side of an equation.	$x-5=19$ $x-5+5=19+5$ $x=24$
Multiplication property of equality.	The process of multiplying/dividing a constant from each side of an equation.	$5x=15$ $\frac{5x}{5}=\frac{15}{5}$ $x=3$
Formula	An equation that describes a relationship between quantities.	$P=2L+2W$ is a formula for perimeter of a rectangle.
Linear inequality in one variable	An inequality that can be written as $ax+b<c$	$2x-3<6$
Intersection	Inequality that contains and or $\cap$	$x<2$ and $x>-4$
Union	Inequality that contains or, or $\cup$	$x>2$ or $x<-4$

Name: **Date:**

Instructor: **Section:**

Practice Test A

Solve each equation.

1. $2x+7=-25$

1. ____________________

2. $-5(x-4)+3(x+3)=12$

2. ____________________

3. $\frac{2x}{3}-\frac{1}{6}=\frac{x}{4}$

3. ____________________

4. $2x-13=4(x-1)-2x$

4. ____________________

5. $|x-2|=5$

5. ____________________

6. $|x|=-9$

6. ____________________

7. $\frac{x-5}{2}-\frac{x+6}{5}=\frac{x}{8}$

7. ____________________

Name: **Date:**

Instructor: **Section:**

8. $|2x-8|=|3x+1|$

8. ____________________

9. $2x-12=|x-6|$

9. ____________________

Solve each equation for the specified variable.

10. $P=2L+2W$; for L

10. ____________________

11. $T=rs^2-prs$; for r

11. ____________________

12. $A=\frac{1}{2}bh$; for b.

12. ____________________

Solve each inequality. Write your answer in interval notation.

13. $x+5>3$

13. ____________________

14. $2x-5<4x+6$

14. ____________________

15. $2(x+3)-4x\geq 13-x$

15. ____________________

Name: **Date:**

Instructor: **Section:**

16. $\frac{2}{3}x - 4 < 6$

16. ____________________

17. $|x-5| \le 3$

17. ____________________

18. $x < 5$ and $x > -2$

18. ____________________

Solve.

19. Find four consecutive odd integers if three times the second number is 15 greater than the fourth integer doubled.

19. ____________________

20. If $A = \{x \mid x \text{ is an even number}\}$ and $B = \{x \mid 15 < x \le 24\}$. Find $A \cap B$

20. ____________________

Name: **Date:**

Instructor: **Section:**

Practice Test B

Solve each equation.

1. $11x - 21 = 34$

1. ______________________

a. $13/11$ b. 5

c. -5 d. $-13/11$

2. $2(x+1) - (4x-5) = -2(x-4) - 1$

2. ______________________

a. $3/2$ b. 4

c. All real numbers d. $\varnothing$

3. $|x-4| = -3$

3. ______________________

a. $7, 1$ b. $-1, -7$

c. $(-\infty, \infty)$ d. $\varnothing$

4. $|2x-6| = |x-5|$

4. ______________________

a. $1, 11/3$ b. $11, 1/3$

c. $-11, -1/3$ d. $-1, -11/3$

Name: **Date:**

Instructor: **Section:**

5. $|4x-9|=2x+1$

5. ______________

a. $-5, -\frac{4}{3}$ b. $5, \frac{4}{3}$

c. $\varnothing$ d. $(-\infty, \infty)$

6. $\dfrac{x-4}{7}=\dfrac{x+2}{9}$

6. ______________

a. 11 b. -25

c. 25 d. -11

7. $0.05x-0.10=0.25$

7. ______________

a. 7 b. 3

c. -15 d. 0.5

8. $|x|=9-x$

8. ______________

a. $\frac{9}{2}$ b. $-\frac{9}{2}$

c. $(-\infty, \infty)$ d. $\varnothing$

9. $17-|x+3|=9$

9. ______________

a. $-29, 23$ b. $-11, 5$

c. $\varnothing$ d. $(-\infty, \infty)$

Name: **Date:**

Instructor: **Section:**

Solve each equation for the specified variable.

10. $y = mx + b$; for *m*.

10. ____________________

a. $m = \frac{y}{x} - b$

b. $m = y - bx$

c. $m = y - \frac{b}{x}$

d. $m = \frac{y-b}{x}$

11. $S = 2lw + 2lh + 2wh$; for *w*.

11. ____________________

a. $w = \frac{S - 2lh}{2l + 2h}$

b. $w = \frac{S - lh}{l + h}$

c. $w = \frac{S}{2l} + \frac{l}{h}$

d. $w = S - \frac{lh}{l + h}$

12. $C = 2\ \ r^2$; for π.

12. ____________________

a. $\ \ = C - 2r^2$

b. $\ \ = \frac{Cr^2}{2}$

c. $\ \ = \frac{C}{2r^2}$

d. $\ \ = \frac{2C}{r^2}$

Solve each inequality. Write your answer in interval notation.

13. $x - 1 < -8$

13. ____________________

a. $(-\infty, 9)$

b. $(9, \infty)$

c. $(-\infty, -7)$

d. $(-7, \infty)$

Name: **Date:**

Instructor: **Section:**

14. $x - 3 < 5(x+1)$

14. ____________________

a. $(2, \infty)$ b. $\left(\frac{1}{2}, \infty\right)$

c. $\left(-\frac{1}{2}, \infty\right)$ d. $(-2, \infty)$

15. $4(x+2) - 6(x-4) < 8x$

15. ____________________

a. $\left(-\frac{8}{5}, \infty\right)$ b. $\left(\frac{16}{5}, \infty\right)$

c. $\left(-\infty, \frac{16}{5}\right)$ d. $\left(-\infty, -\frac{16}{3}\right)$

16. $\frac{1}{2}(6x - 18) > 12$

16. ____________________

a. $(7, \infty)$ b. $(-7, \infty)$

c. $(1, \infty)$ d. $(-1, \infty)$

17. $|2x + 7| > 9$

17. ____________________

a. $(-2, 8)$ b. $(-8, 1)$

c. $(-\infty, -2) \cup (8, \infty)$ d. $(-\infty, -8) \cup (1, \infty)$

18. $x \geq 25$ or $x < 19$

18. ____________________

a. $(19, 25)$ b. $(-\infty, 19) \cup [25, \infty)$

c. $(19, 25]$ d. $(-\infty, 25) \cup (19, \infty)$

Name: **Date:**

Instructor: **Section:**

Solve.

19. The perimeter of a triangle is 53 meters, find the length of the sides if the first side is 3 more than the second side, and the third side is twice the first side.

19. ______________________

a. 14, 11, 28

b. 15, 19, 7

c. 12, 16, 6

d. 10, 13, 5

20. Find the union on $A = \{2,5,8,9\}$ and $B = \{1,6,8,11\}$.

20. ______________________

a. $\{5,6,8,9\}$

b. $\{8\}$

c. $\varnothing$

d. $\{1,2,5,6,8,9,11\}$

Name: **Date:**

Instructor: **Section:**

Chapter 3 Graphs and Functions
Section 3.1 Graphing Equations

Learning Objectives

1. Plot ordered pairs.
2. Determine whether an ordered pair is a solution to an equation in two variables.
3. Graph linear equations.
4. Graph nonlinear equations.

Objective 1

In words, describe where the given point will be graphed. State the quadrant in which it will be.

1. $(-10,16)$

1. ____________________

2. $(5,-9)$

2. ____________________

3. $(7,19)$

3. ____________________

4. $(-15,-18)$

4. ____________________

Objective 2

Determine whether each ordered pair is a solution of the given equation.

5. $2x+3y=7;\quad (2,-1)$

5. ____________________

6. $-6x+5y=-6;\quad (1,0),\left(2,\frac{6}{5}\right)$

6. ____________________

Name: **Date:**

Instructor: **Section:**

7. $y = 2x^3 - 4x^2 + 3x - 6;\quad (-1, -15)$

7. ____________________

8. $y = |x + 2|;\quad (-6, -4)$

8. ____________________

Objective 3

Graph the following equations by finding and plotting ordered pair solutions.

9. $2x - 3y = 6$

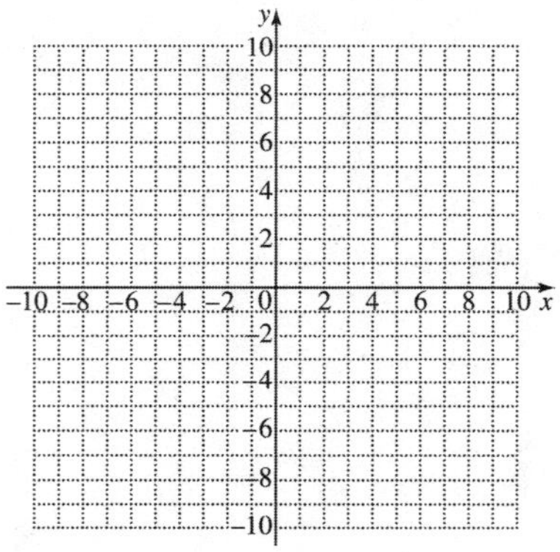

10. $y = \frac{3}{4}x - 8$

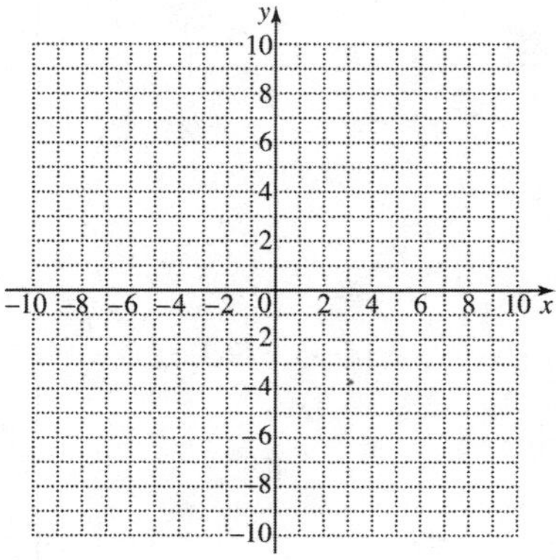

Name: **Date:**

Instructor: **Section:**

11. $x = 2y - 6$

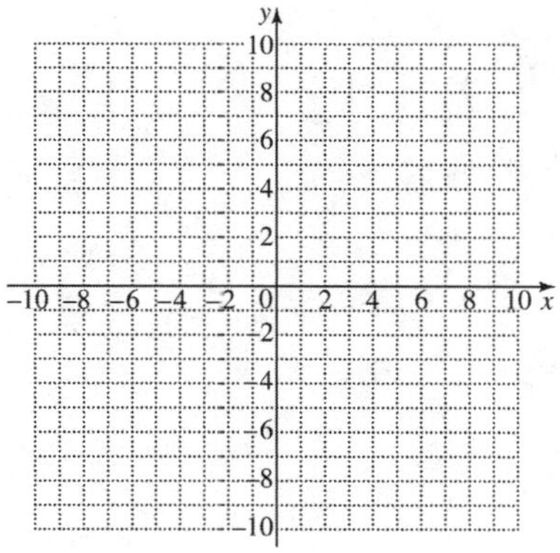

12. $y = -3x$

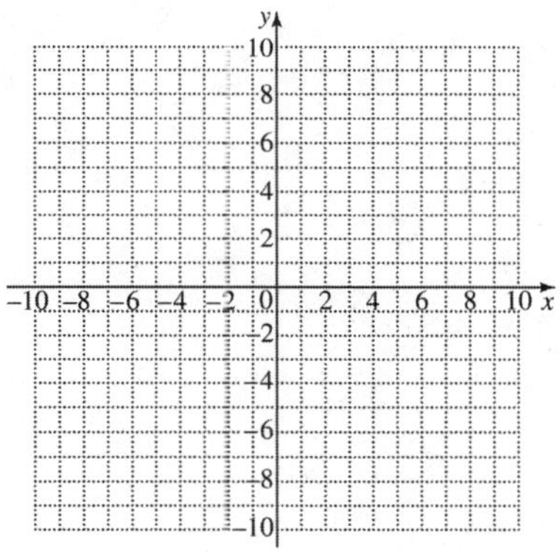

Name: **Date:**

Instructor: **Section:**

Objective 4

Graph the following equation by finding points and plotting ordered pair solutions.

13. $y = 2|x|$

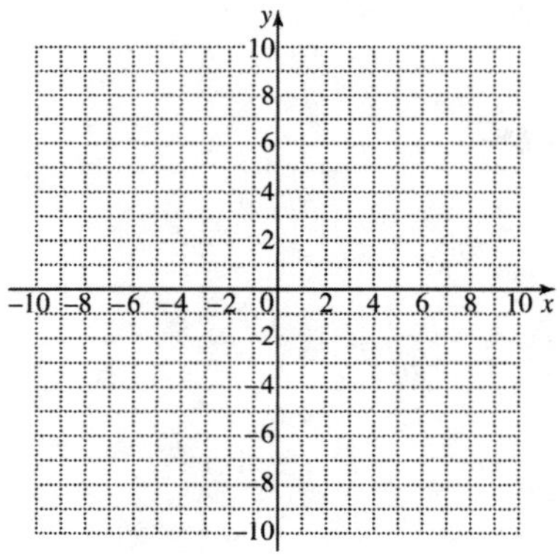

14. $y = x^2 - 4$

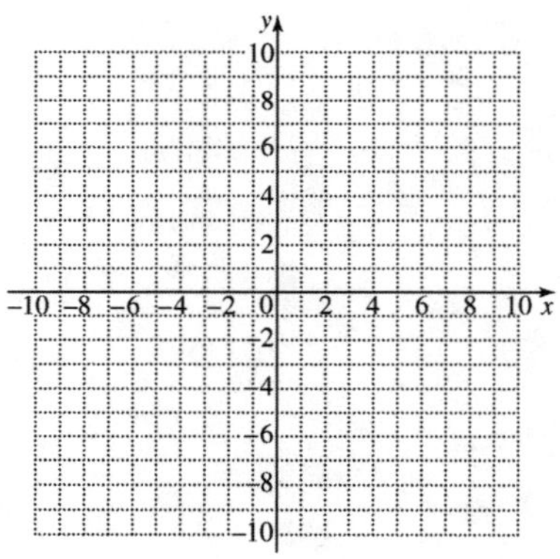

Name: **Date:**

Instructor: **Section:**

15. $y = x^3 - 7$

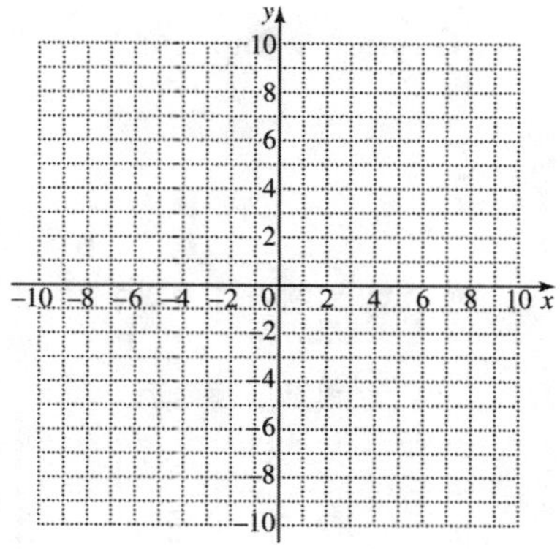

16. $y = |x + 3|$

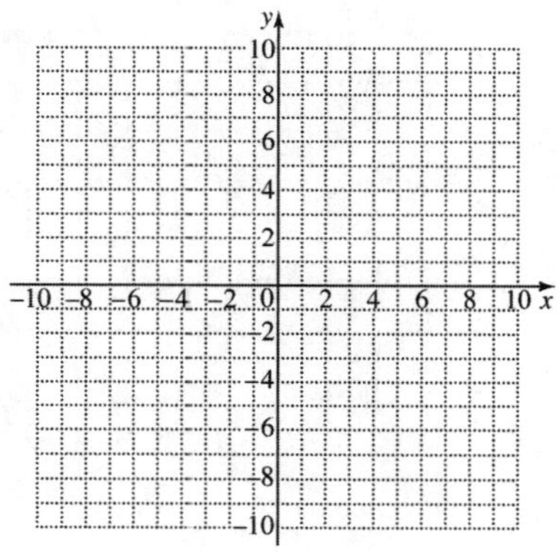

Name: **Date:**

Instructor: **Section:**

Concept Extension

17. Graph $y = x^2$ and $y = (x-2)^2 + 5$ on the same rectangular coordinate system. Explain what the difference happens to be between the two graphs.

17. ____________________

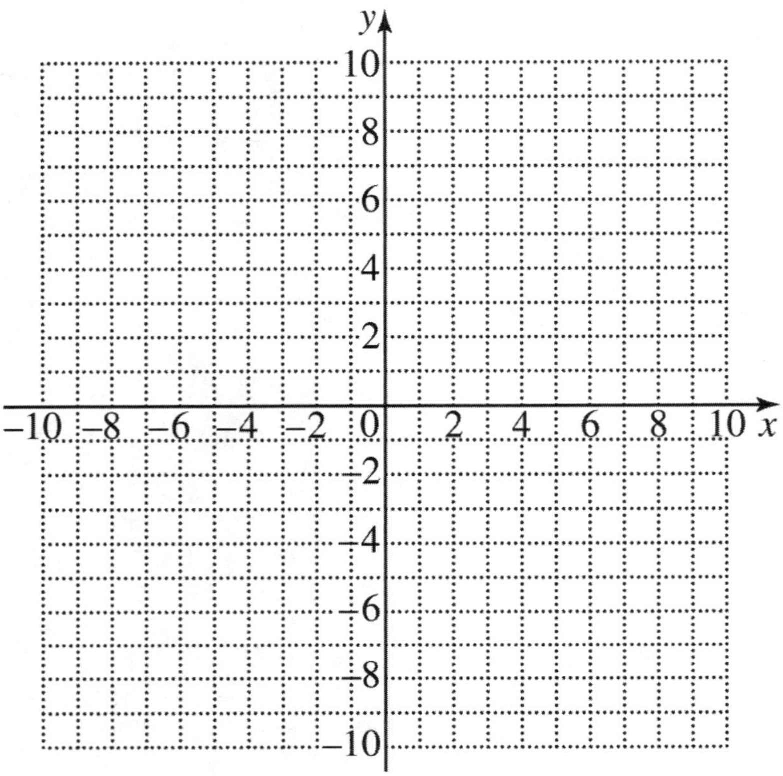

Name: **Date:**

Instructor: **Section:**

Section 3.2 Introduction to Functions

Learning Objectives

1. Define relation, domain, and range.
2. Identify functions.
3. Use the vertical line test for functions.
4. Find the domain and range of a function.
5. Use function notation.

Vocabulary

Use the choices to complete each statement.

Domain	**Function**	**Horizontal**
Range	**Relation**	**Vertical**
x	**y**	

1. A ________________ is a set of ordered pairs that for every x value there is only one y value.
2. The ___________________ of the function is the x- values.
3. To find the x-intercept, you would let ________ = 0.
4. To determine if a graph is a function, it must pass the ________________ line test.
5. A set of ordered pairs is called a _____________________.
6. The _________________ of a function is the y-values.
7. To find the y-intercept, you would let ___________ = 0

.

Objective 1

Find the domain and range of the function.

8. $\{(-2,4),(8,-9),(-4,6),(-2,8)\}$

8. ______________________

9. $\{(1,7),(-7,3),(0,-6),(12,15)\}$

9. ______________________

10. $\{(-12,16),(33,45),(-72,99),(110,-114)\}$

10. ___________________

Name: **Date:**

Instructor: **Section:**

Objective 2

Decide whether each is a function.

11. $y = x + 1$

11. ____________________

12. $\{(12,7),(-17,3),(1,-6),(12,-5)\}$

12. ____________________

13. $y = x^2 - 9$

13. ____________________

14. $x = y^4 + 5$

14. ____________________

Objective 3

Use the vertical line test to determine whether each graph is a function.

15.

15. ____________________

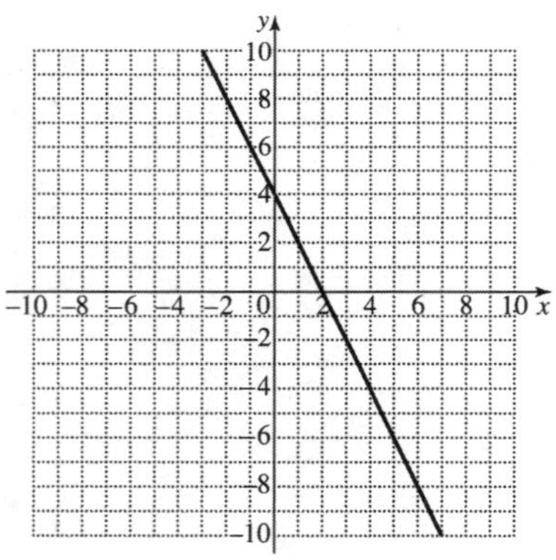

Name: **Date:**

Instructor: **Section:**

16.

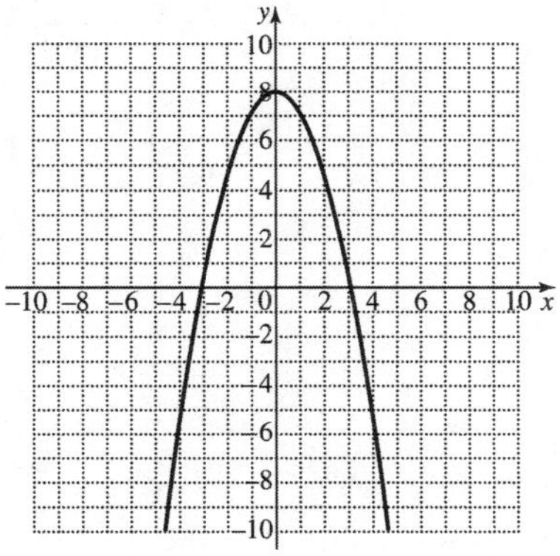

16. ____________________

17.

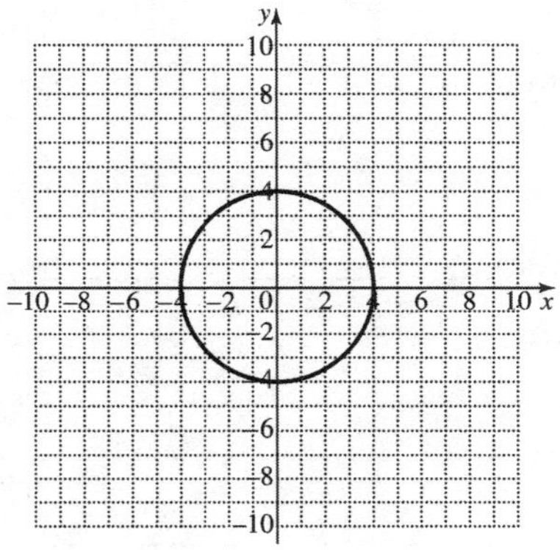

17. ____________________

Objective 4

Find the domain and range of each relation.

18.

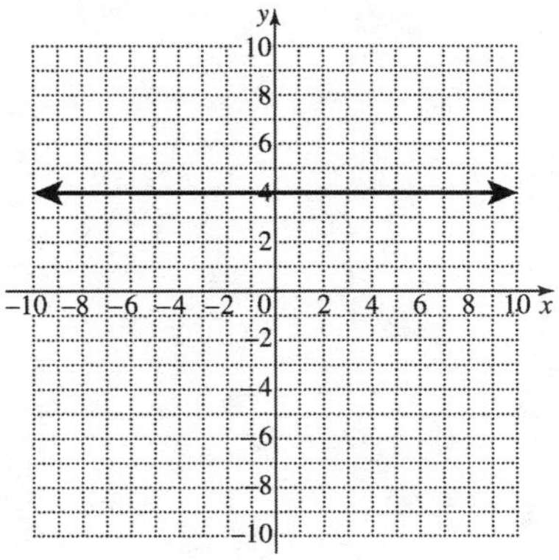

18. ____________________

Name: **Date:**

Instructor: **Section:**

19.

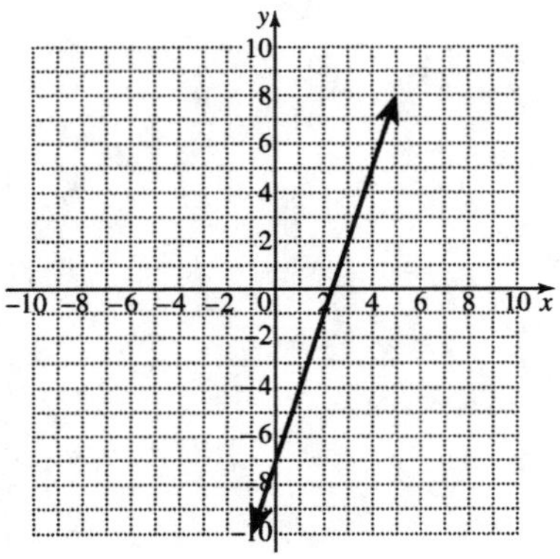

19. ____________________

Objective 5

If $f(x) = 3x + 3$ and $g(x) = 4x^2 - 6x + 3$, and $h(x) = 5x^2 - 7$, find the following.

20. $h(-3)$

20. ____________________

21. $f(7)$

21. ____________________

22. $g(-5)$

22. ____________________

Name:

Instructor:

Date:

Section:

Concept Extension.

Given the function $f(x) = 3x - 4$.

23. Find $f(a)$ and $f(a-h)$.

23. ____________________

24. Using the function in number 23, find $\frac{f(a) - f(a-h)}{h}$.

24. ____________________

Name: **Date:**

Instructor: **Section:**

Section 3.3 Graphing Linear Functions

Learning Objectives

1. Graph linear function
2. Graph linear functions by finding intercepts.
3. Graph vertical and horizontal lines.

Vocabulary

Use the choices to complete each statement.

Horizontal	**Linear**	**Vertical**
b	**m**	**x**
y		

1. The equation $y-8=4$ is a ________________________ line.

2. A __________________ function's equation can be written in the form $f(x)=mx+b$.

3. The equation $x=-5$ is a ____________________ line.

4. In the equation $y=mx+b$, the _________ represents the slope, and the ____________ represents the ________ -intercept.

Objective 1

Graph each linear function.

5. $f(x)=-2x+3$

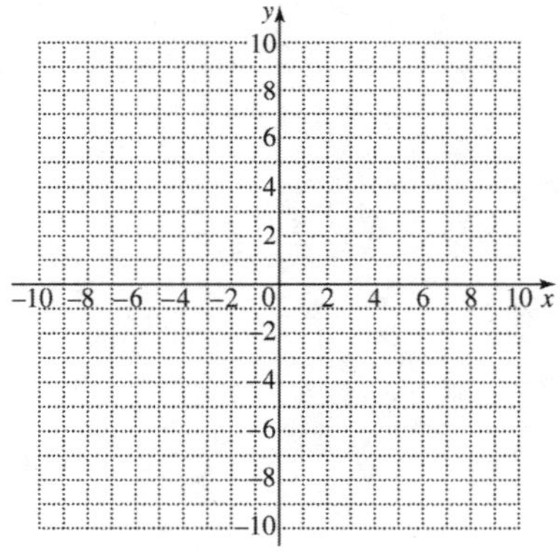

Name: **Date:**

Instructor: **Section:**

6. $f(x) = -x - 4$

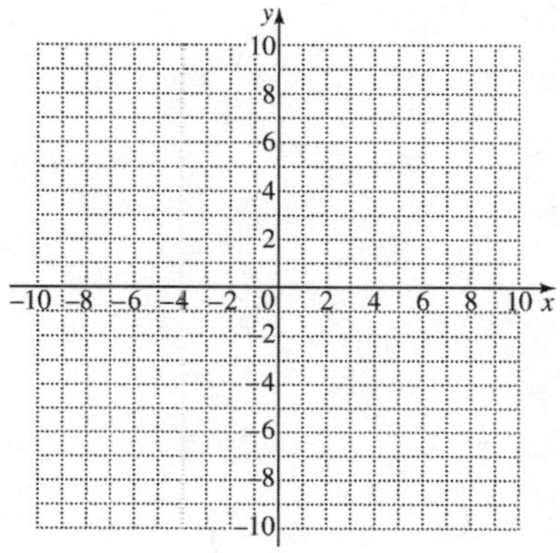

7. $f(x) = -\frac{1}{3}x$

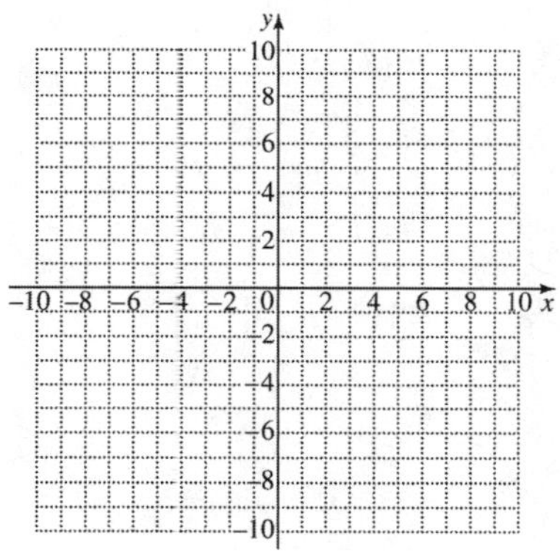

8. $f(x) = 0.25x - 6$

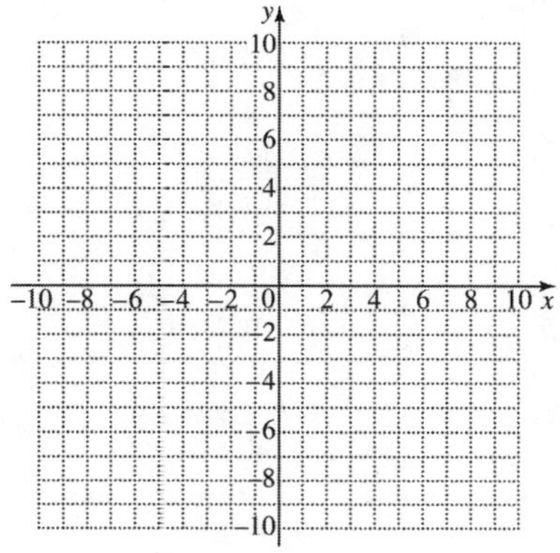

Name: **Date:**

Instructor: **Section:**

Objective 2

Graph each linear function by finding x- and y- intercepts.

9. $3x - 2y = 12$

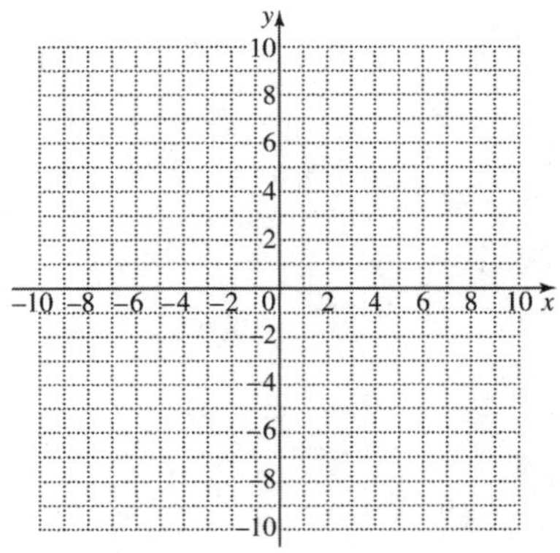

10. $x - y = -6$

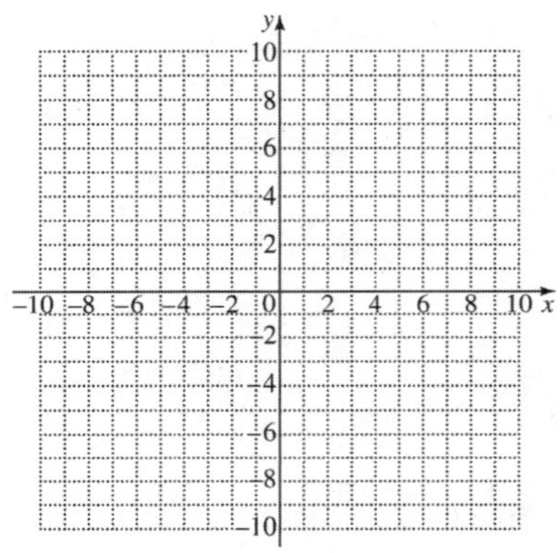

11. $2x - 3y = -9$

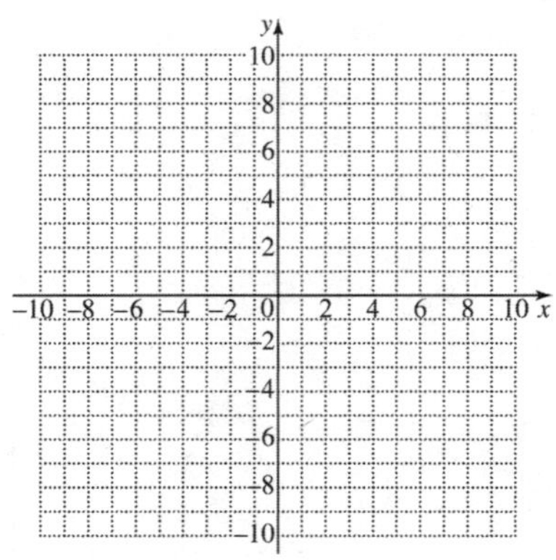

Name: **Date:**

Instructor: **Section:**

12. $-y+3x=9$

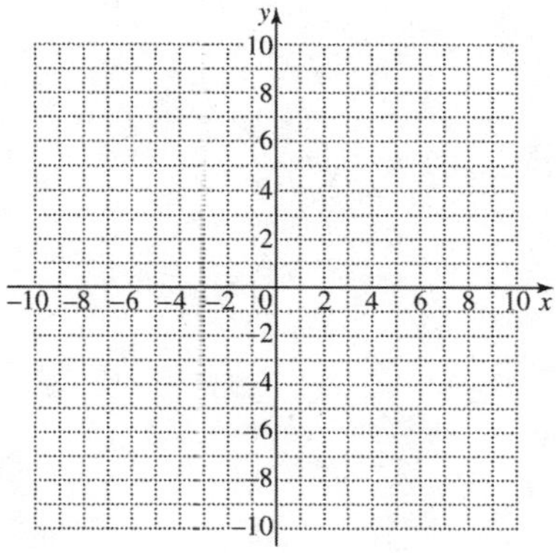

Objective 3

Graph each linear equation.

13. $x=-1$

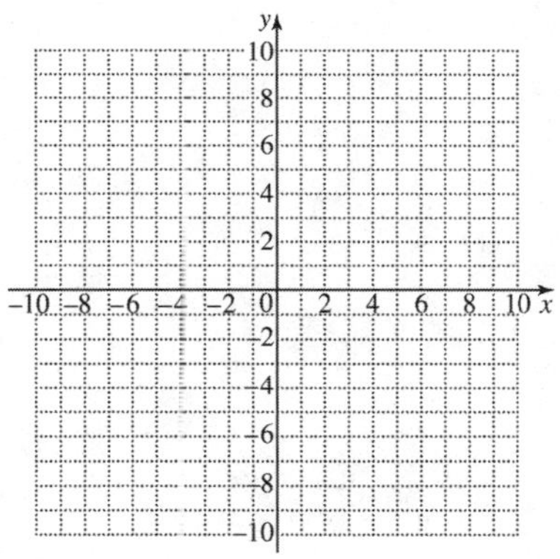

Name: **Date:**

Instructor: **Section:**

14. $y - 2 = -3$

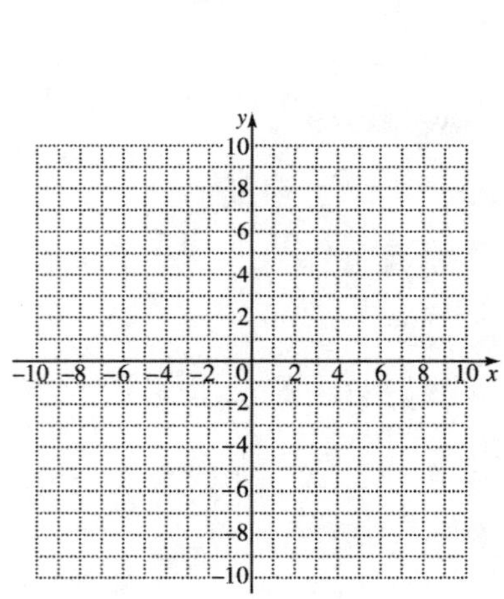

15. $-5x = 15$

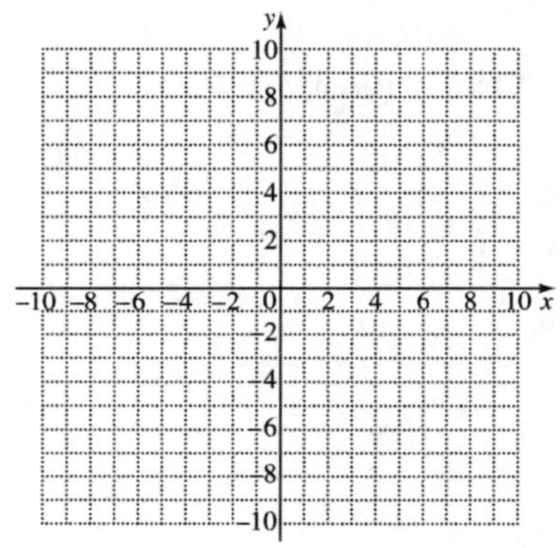

16. $-y = -6$

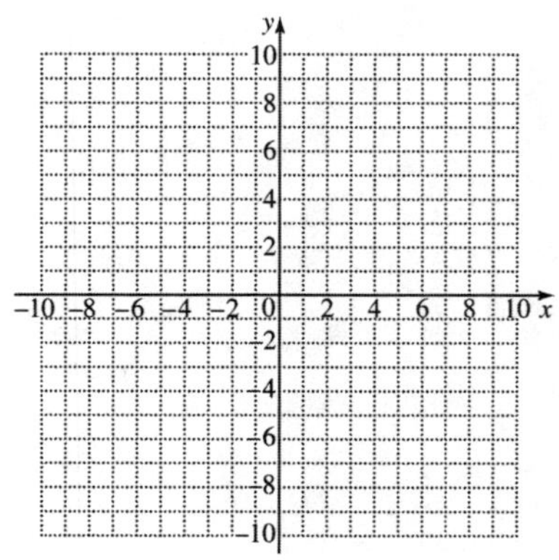

Name: **Date:**

Instructor: **Section:**

Concept Extension

17. Given the equation $f(x) = 7$, what is the coefficient in front of the x? What is the slope of this line?

17. ______________________

Name: **Date:**

Instructor: **Section:**

Section 3.4 The Slope of a Line

Learning Objectives

1. Find the slope of a line given two points on the line.
2. Find the slope of a line given the equation of a line.
3. Interpret the slope-intercept form in an application
4. Find the slopes of horizontal and vertical lines.
5. Compare the slopes of parallel and perpendicular lines.

Vocabulary

Use the choices to complete each statement.

Different	**Equal**	**Perpendicular**
Slope	**Slope-Intercept**	**Undefined**
0	**1**	

1. Two parallel lines will have __________ slopes, but ______________ intercepts.
2. The slope of a vertical line is ______________.
3. The rate of change in which a line increases or decreases, is the __________________.
4. If the slopes of two lines are opposite reciprocals, they are said to be _____________________.
5. The form $y = mx + b$ is called ______________________ form.
6. The slope of the line $y = 1$ is __________________.

Objective 1

Find the slope of the line that runs through the given points.

7. $(-2,-6), (4,-4)$

7. ______________________

8. $(-1,0), (-4,-9)$

8. ______________________

9. $(7,-2), (7,9)$

9. ______________________

Name: **Date:**

Instructor: **Section:**

Objective 2

Find the slope of each line.

10. $y = \frac{1}{5}x - 9$

10. ____________________

11. $4x - 7y = 10$

11. ____________________

12. $f(x) = -5x$

12. ____________________

Objective 3

Solve.

13. With wireless Internet (WiFi) gaining popularity, the number of public wireless Internet access points (in thousands) is projected to grow from 2003 to 2008 according to the equation $-66x + 2y = 84$ where x is the number of years after 2003.

    a. Find the slope and y-intercept of the linear equation.

    b. What does the slope mean in this context?

    c. What does the y-intercept mean in this context?

13a. ____________________

13b. ____________________

13c. ____________________

Name: **Date:**

Instructor: **Section:**

14. The cost C, in dollars, of renting a convertible for a day is given by the equation $C(x) = 0.25x + 64$, where x is the number of miles driven.

    a. Find the slope and y-intercept of the linear equation.

    b. What does the slope mean in this context.

    c. What does the y-intercept mean in this context.

14a. ______________________

14b. ______________________

14c. ______________________

Objective 4

Find the slope of each line.

15. $x = -5$

15. ______________________

16. $y = 13$

16. ______________________

17. $-3x = 12$

17. ______________________

18. $y + 6 = -12$

18. ______________________

Name: **Date:**

Instructor: **Section:**

Objective 5

Determine whether the lines are parallel, perpendicular, or neither.

19.

$f(x) = -3x + 6$

$g(x) = 3x + 5$

19. ____________________

20.

$y = \frac{1}{3}x - 4$

$-x + 3y = 7$

20. ____________________

21.

$2x + 3y = 9$

$3x - 2y = 9$

21. ____________________

Concept Extension

22. Find the slope of the line that is perpendicular to $4x - 7y = 14$.

22. ____________________

23. Find the slope of the line that is parallel to $12x - 15y = 18$.

23. ____________________

Name: **Date:**

Instructor: **Section:**

Section 3.5 Equations of Lines

Learning Objectives

1. Use the slope-intercept form to write the equation of a line.
2. Graph a line using its slope and *y*-intercept.
3. Use the point-slope form to write the equation of a line.
4. Write equations of vertical and horizontal lines.
5. Find equations of parallel and perpendicular lines.

Objective 1

Use the slope-intercept form of the linear equation to write the equation of each line with the given slope and *y*-intercept.

1. Slope $= -2$; *y*-intercept $(0, -4)$

1. ______________________

2. Slope $= \frac{2}{3}$; *y*-intercept $(0, 6)$

2. ______________________

3. Slope $= -\frac{3}{5}$; *y*-intercept $\left(0, -\frac{7}{9}\right)$

3. ______________________

Name: **Date:**

Instructor: **Section:**

Objective 2

Graph each linear equation using slope-intercept form.

4. $y = \frac{2}{5}x - 8$

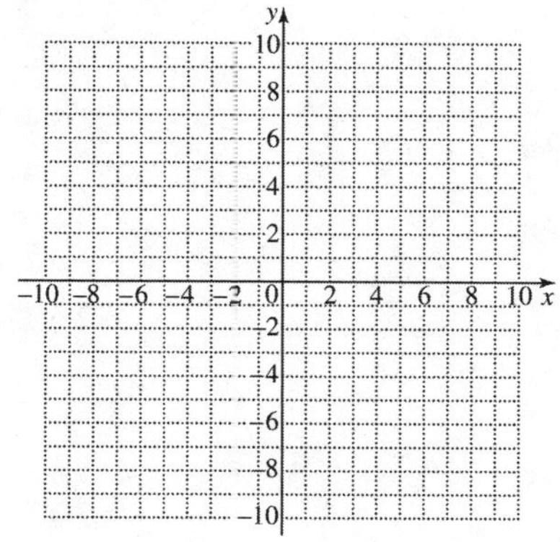

5. $4x - 3y = 8$

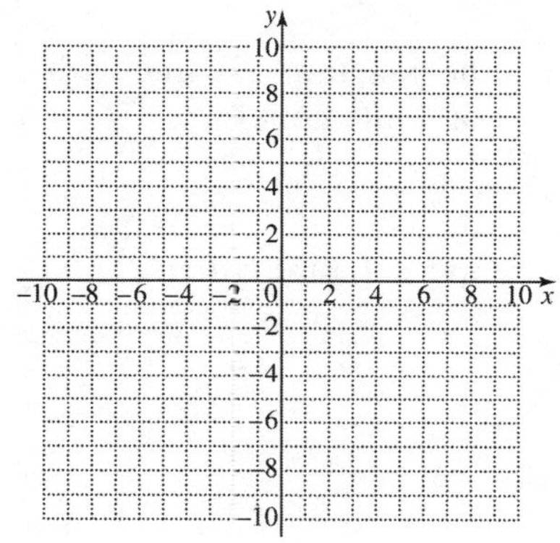

6. $-3x - 6y = 12$

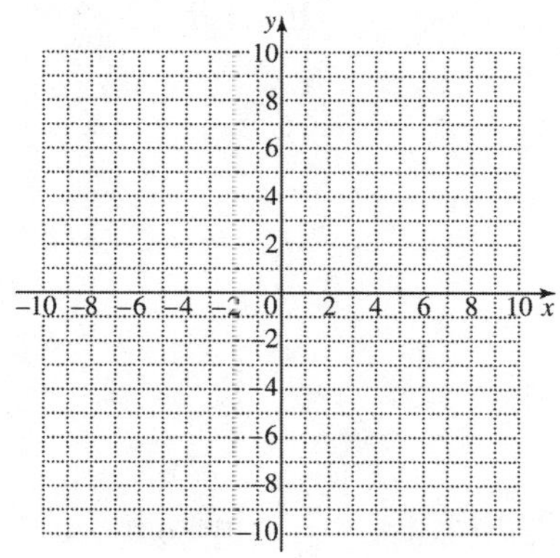

Name: **Date:**

Instructor: **Section:**

7. $x = 3y - 9$

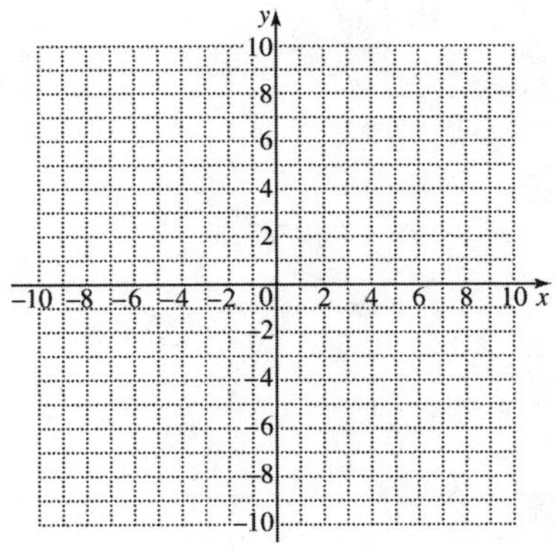

Objective 3

Find the equation of a line with the given slope and containing the given point. Write the equation in slope-intercept form.

8. Slope = 3; through $(1, 2)$

8. ______________________

9. Slope = $\frac{1}{4}$; through $(8, -3)$

9. ______________________

10. Slope = $-\frac{2}{5}$; through $(3, -4)$

10. ______________________

Name: **Date:**

Instructor: **Section:**

11. Slope = -7; through $\left(\frac{1}{2}, \frac{3}{5}\right)$

11. ______________________

Objective 4

Find the equation of each line.

12. Slope = 0; through $(-1, -5)$

12. ______________________

13. Vertical line; through $(2, 6)$

13. ______________________

14. Undefined slope; through $\left(-5, -\frac{2}{3}\right)$

14. ______________________

15. Horizontal line; through $\left(\frac{2}{5}, -\frac{1}{3}\right)$

15. ______________________

Objective 5

Find the equation of each line. Write the equation using function notation.

16. Through $(2, -5)$; perpendicular to $3y = x - 6$

16. ______________________

Name: **Date:**

Instructor: **Section:**

17. Through $(1,-3)$; parallel to $x-3y=7$

17. ____________________

18. Through $(3,-5)$; parallel to $y=-3$

18. ____________________

19. Through $(0,-23)$; perpendicular to $2x-3y=8$

19. ____________________

Concept Extension

Find the slope of the line through the following points. Use function notation to write the equation.

20. $(-3,-8),(-6,-9)$

20. ____________________

21. $\left(\frac{3}{4},-\frac{2}{5}\right),\left(-\frac{1}{5},\frac{1}{4}\right)$

21. ____________________

Name: **Date:**

Instructor: **Section:**

Section 3.6 Graphing Piecewise-Defined Functions and Shifting and Reflecting Graphs of Functions

Learning Objectives

1. Graph piecewise-defined functions.
2. Vertical and horizontal shifts.
3. Reflect graphs.

Objective 1

Graph each piecewise-defined function.

1. $f(x)=\begin{cases} x+3 & \text{if } x>1 \\ -2x+4 & \text{if } x\le -1 \end{cases}$

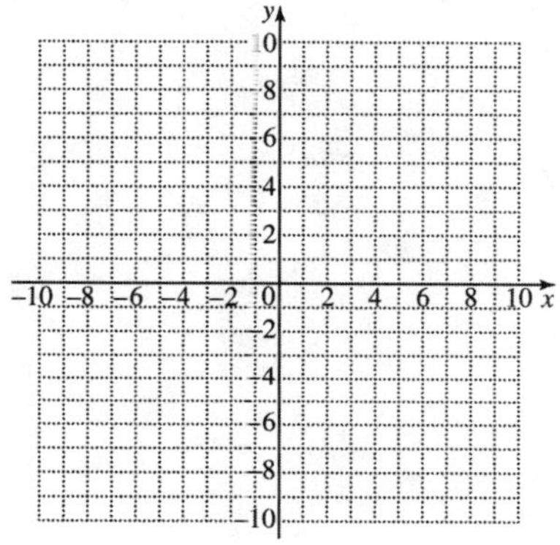

2. $f(x)=\begin{cases} 2x-1 & \text{if } x>2 \\ 3x+5 & \text{if } x\le -2 \end{cases}$

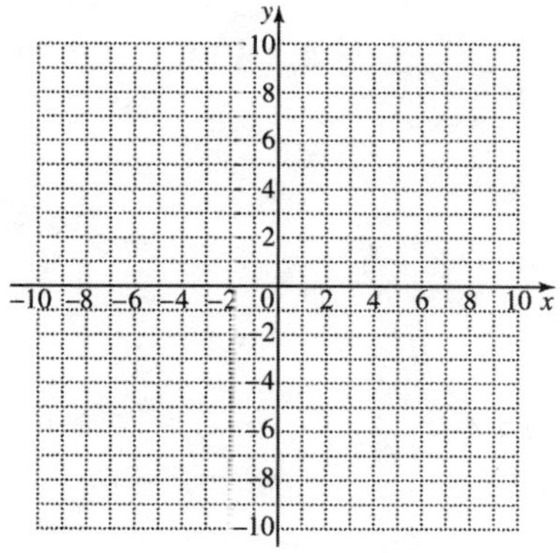

Name: **Date:**

Instructor: **Section:**

3. $f(x) = \begin{cases} -2 & \text{if } x \le -3 \\ 2x - 6 & \text{if } x > 2 \end{cases}$

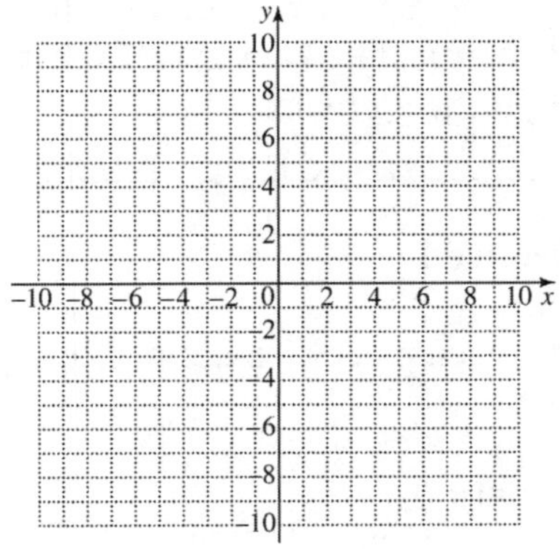

4. $f(x) = \begin{cases} x - 5 & \text{if } x > 4 \\ 3x - 4 & \text{if } x \le -1 \end{cases}$

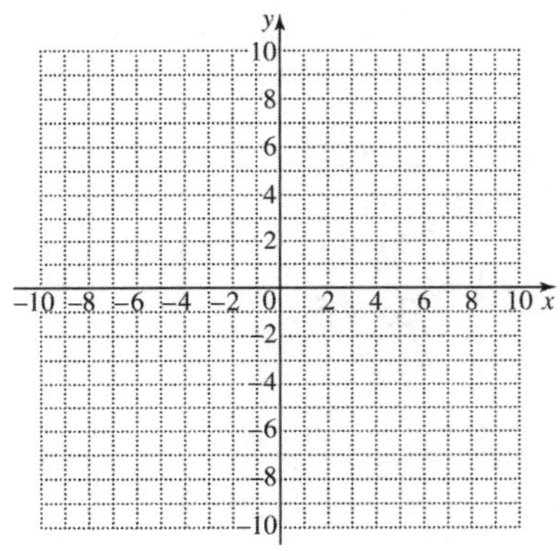

Name: **Date:**

Instructor: **Section:**

Objective 2

Sketch the graph of each function.

5. $f(x) = x^2 - 4$

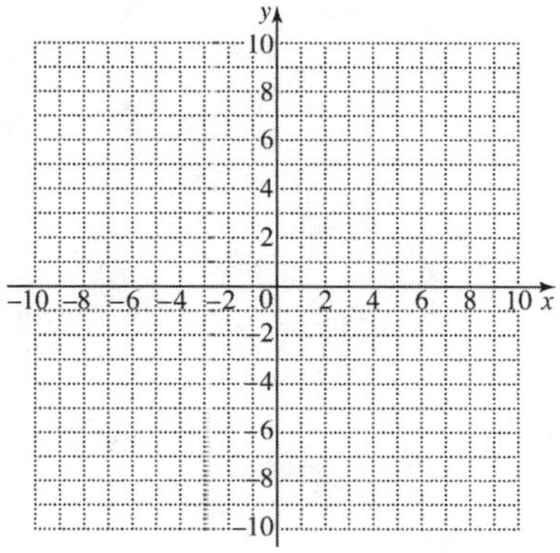

6. $f(x) = |x + 2|$

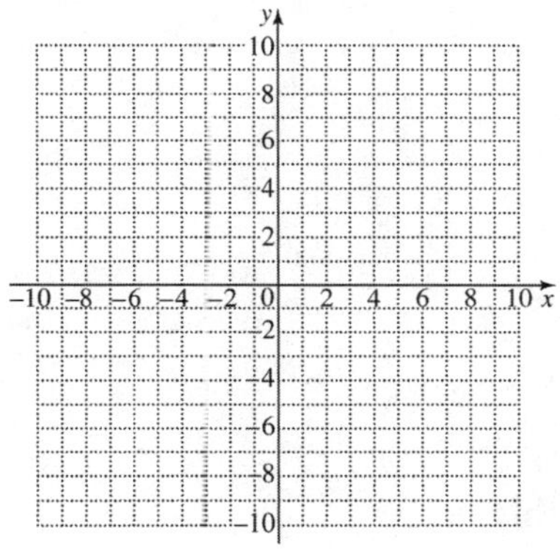

7. $f(x) = (x - 1)^2$

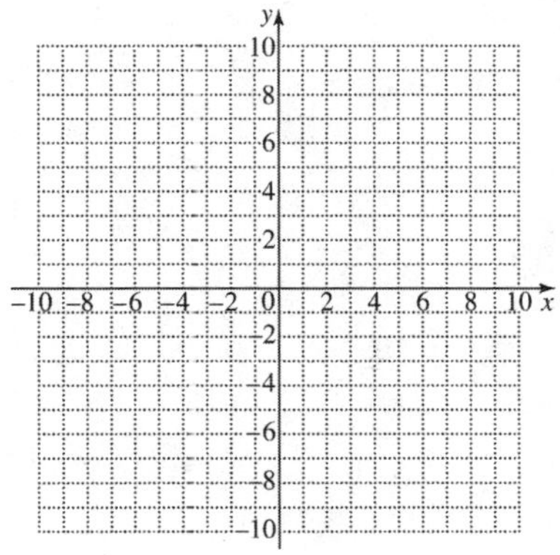

Name: Date:

Instructor: Section:

8. $f(x) = \sqrt{x+1} + 1$

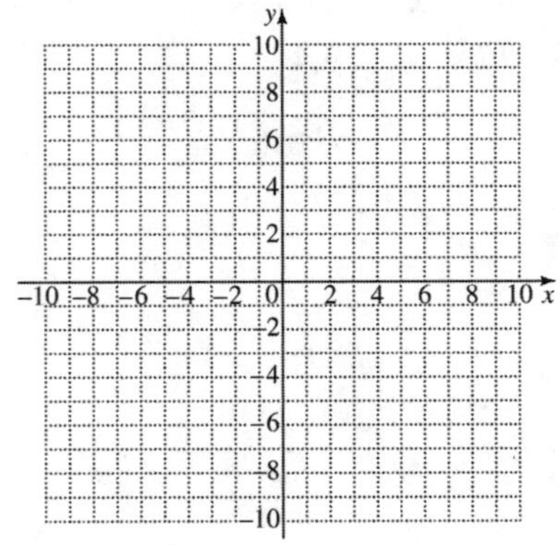

9. $f(x) = |x+1| - 5$

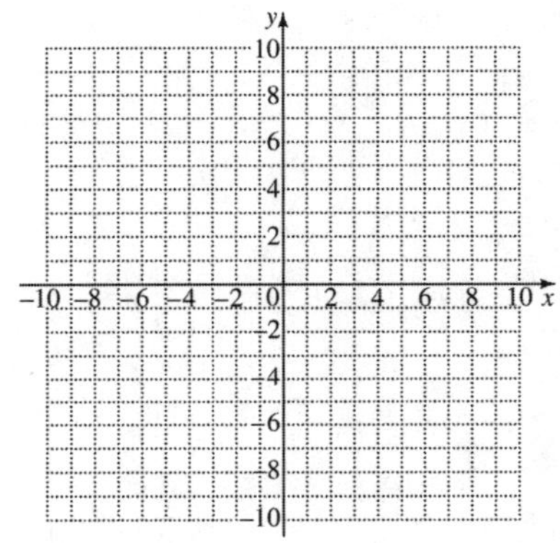

Objective 3

Sketch the graph of each function.

10. $f(x) = -x^2 - 2$

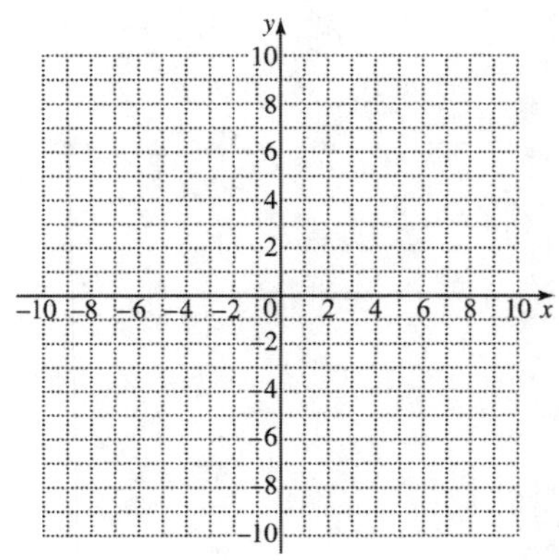

Name: **Date:**

Instructor: **Section:**

11. $f(x) = -|x+1| + 5$

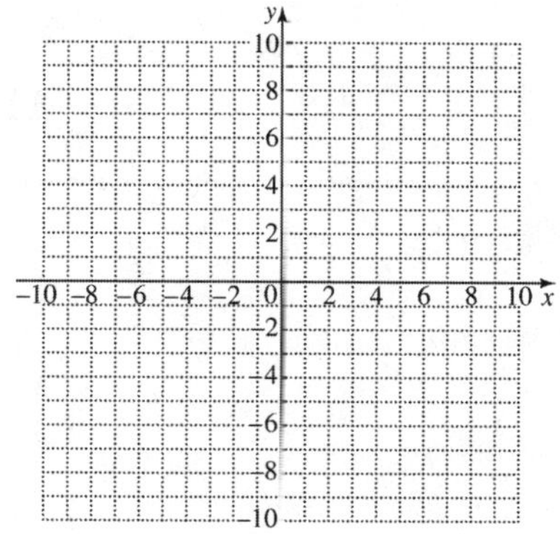

12. $f(x) = -\sqrt{x-2} - 5$

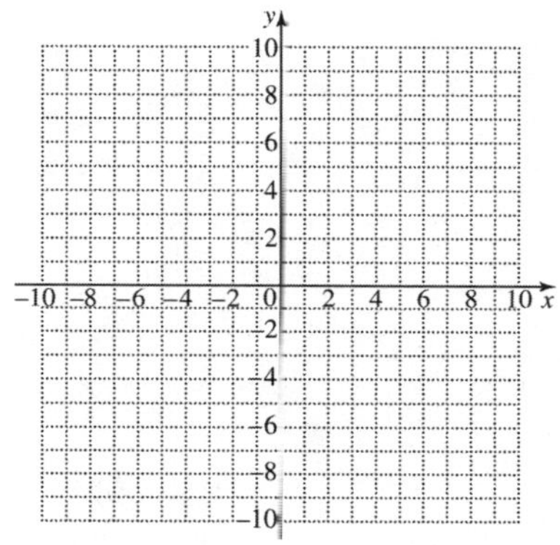

Concept Extension

Graph the following piecewise-defined function.

13. $f(x) = \begin{cases} -2x+6 \text{ if } x > 5 \\ -x+4 \text{ if } -2 \le x < 5 \\ -3x-9 \text{ if } x < -2 \end{cases}$

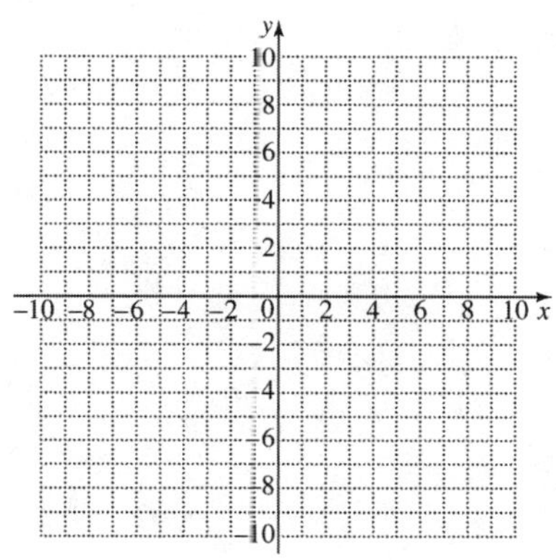

Name: **Date:**

Instructor: **Section:**

Section 3.7 Graphing Linear Inequalities

Learning Objectives

1. Graph linear inequalities.
2. Graph the intersection or union of two linear inequalities.

Objective 1

Graph each inequality.

1. $x > 4$

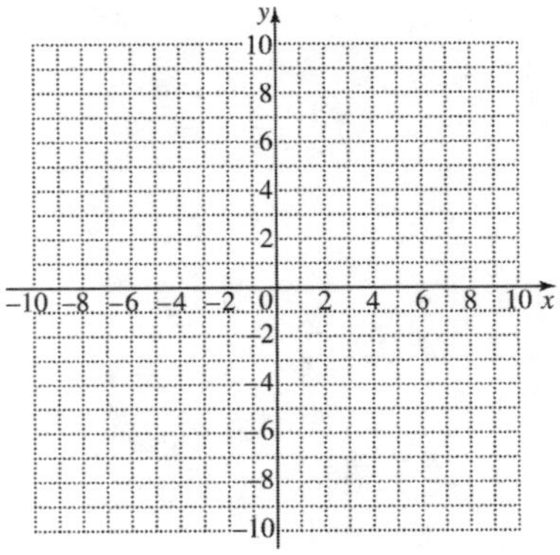

2. $y \le \frac{1}{2}x - 5$

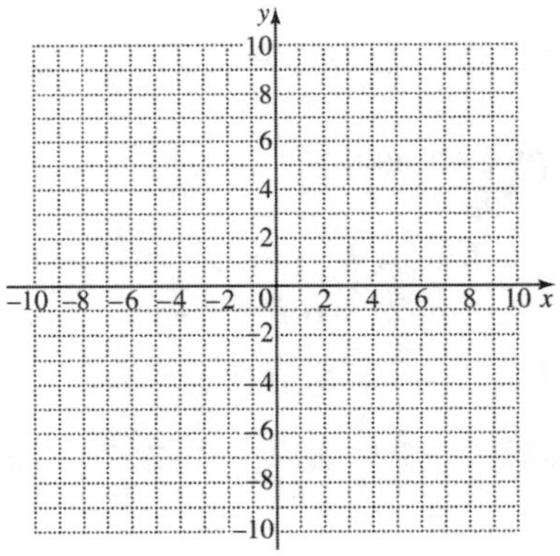

Name: Date:

Instructor: Section:

3. $3x + y > 6$

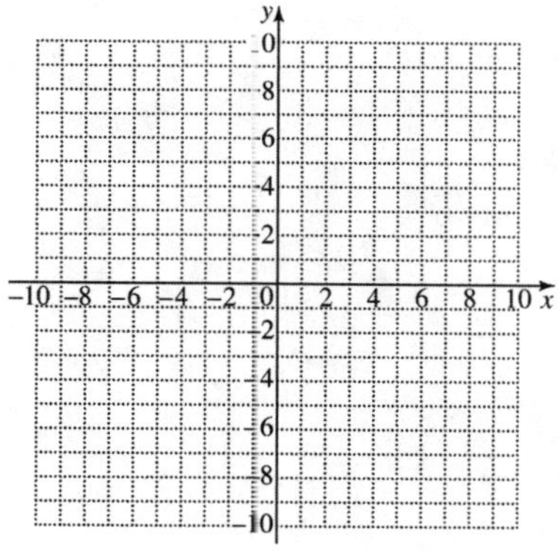

4. $2x - 4y < 16$

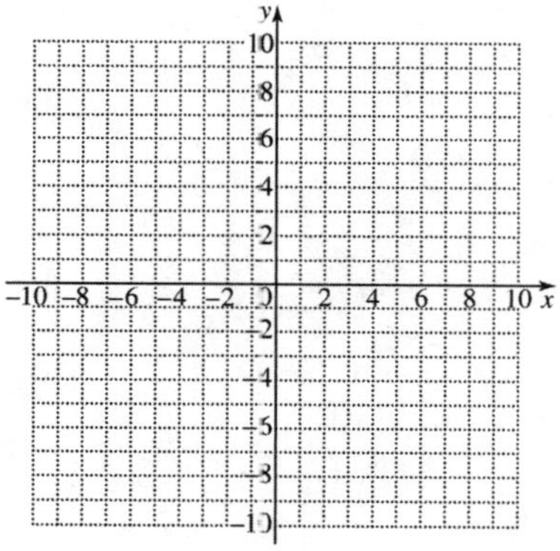

Objective 2

5. Graph the intersection of $x + y \le 1$ and $y \le -1$

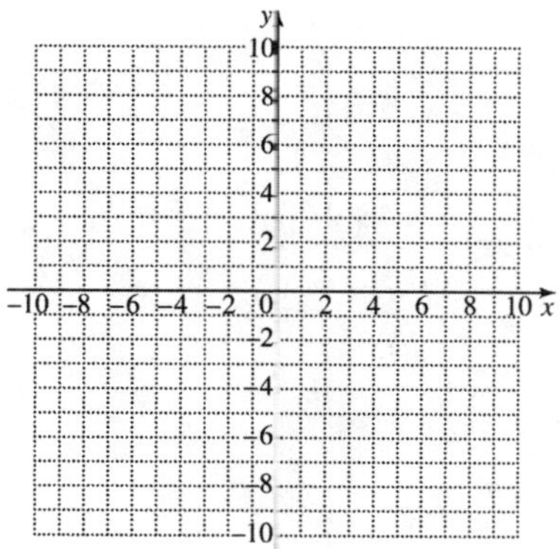

Name:

Instructor:

Date:

Section:

6. Graph the union of $x + y > 4$ and $x > 2$

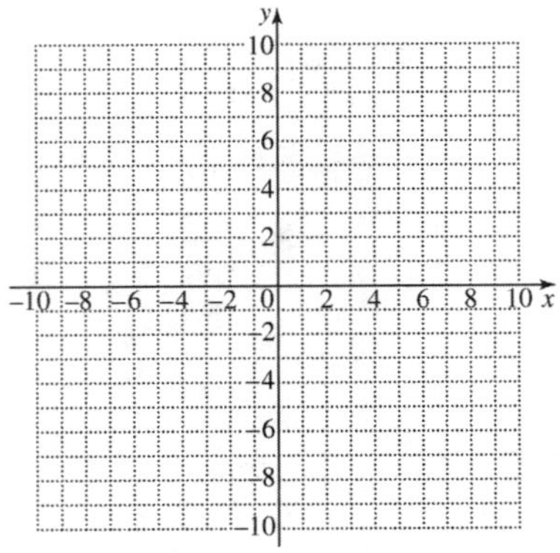

7. Graph the union of $2x + 3y \leq 6$ and $y < 2$

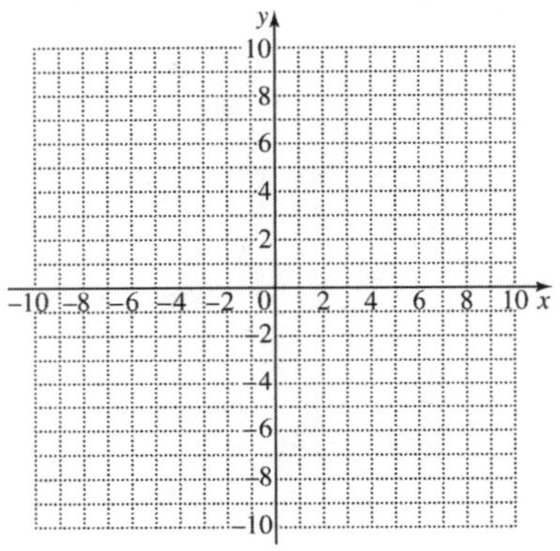

8. Graph the intersection of $x - y < -5$ and $x \geq -3$.

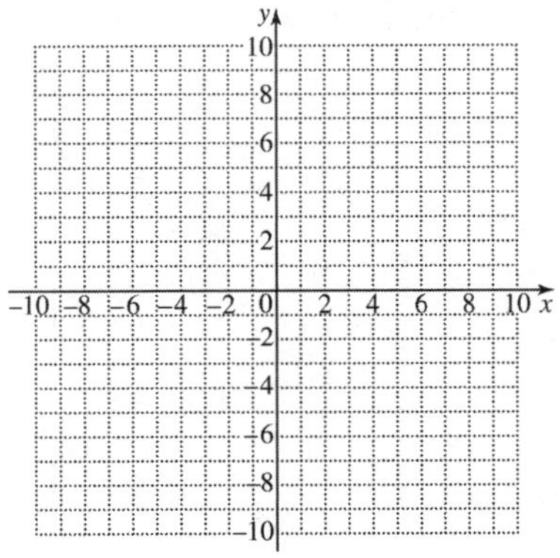

Name: **Date:**

Instructor: **Section:**

Concept Extension

9. Graph the intersection of the following equations:

$$\begin{cases} x - y > 2 \\ y < 2x \\ y > 4 \end{cases}$$

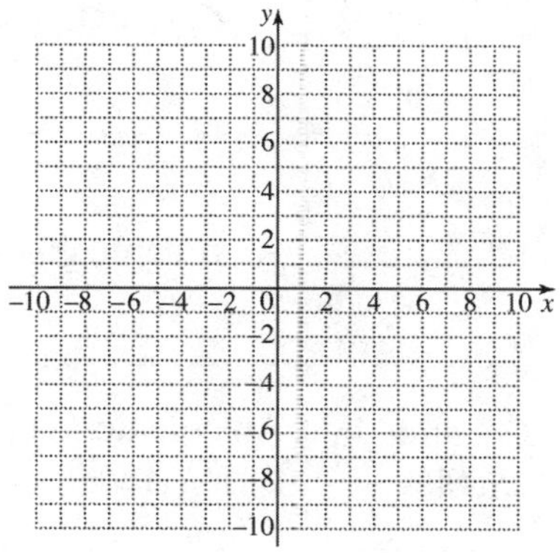

10. Graph the union of the following equations:

$$\begin{cases} x + y < 5 \\ y > 2x - 1 \\ x < -2 \end{cases}$$

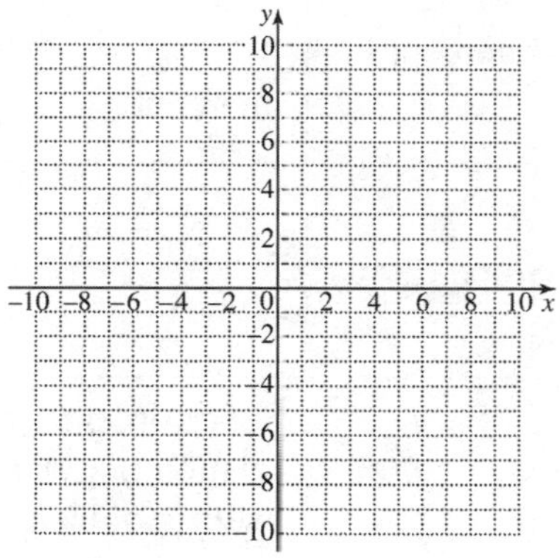

Name: **Date:**

Instructor: **Section:**

Chapter 3 Vocabulary

Vocabulary Word	Definition	Example
Rectangular coordinate system	A plane that contains a vertical (y) and horizontal (x) axes. The intersection of the axes is called the origin.	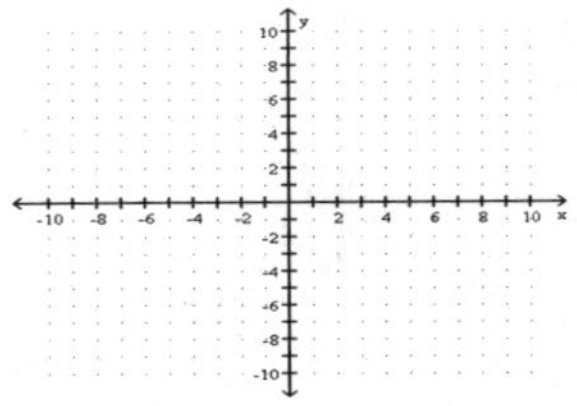
Solution	An ordered pair that makes the equation true.	$y = 3x + 2$ Solution $(-2, -4)$ $-4 = 3(-2) + 2$
Linear Equation in Two Variables	An equation that can be written in standard form: $Ax + By = C$	$5x - 4y = 12$ $y = 2x - 8$
Intercept	A point on the graph where it hits a specific axis.	$y = 2x + 8$ Hits the y axis at 8.
Slope (m)	Rate of change. Steepness of the graph.	$\frac{y_2 - y_1}{x_2 - x_1}$
Slope-Intercept Form.	$y = mx + b$ where m is the slope and b is the y-intercept.	$y = 4x - 9$
Point-Slope Form	$y - y_1 = m(x - x_1)$ where m is the slope and the point (x_1, y_1)	Given m = 2 and $(1,3)$ then $y - 3 = 2(x - 1)$
Relation	Set of ordered pairs. A function is a special relation that the x values do not repeat. For every x-value there is only one y-value.	$\{(1,2)(2,3)(4,5)\}$ Is a function as well.
Domain	Set of all the x-values in a relation.	For the relation above: D={1, 2, 4}
Range	Set of all the y-values in a relation.	For the relation above: R= {2, 3, 5}
$f(x)$	Function notation. A function of x.	$f(x) = 3x + 2$

Name: **Date:**

Instructor: **Section:**

Linear function	A function that can be written as $f(x)=mx+b$.	$f(x)=3x-6$
Vertical Shifts	$g(x)=f(x)+k$; If k positive, graph shifts up; if k is negative, graph shifts down.	$f(x)=x^2+3$; graph is three units higher than $f(x)=x^2$
Horizontal Shifts	$g(x)=f(x-h)$; If h is positive, graph shifts to the right; if h is negative, graph shifts to the left.	$f(x)=\lvert x-4\rvert$; graph is 4 units to the right of $f(x)=\lvert x\rvert$
Linear Inequality in Two Variables	$f(x)<ax+b$ $f(x)\geq ax+b$	$f(x)>2x-6$

Name: **Date:**

Instructor: **Section:**

Practice Test A

1. What quadrant does the point $(-3,5)$ fall in?

1. ______________________

Graph each line.

2. $y = 2x - 3$

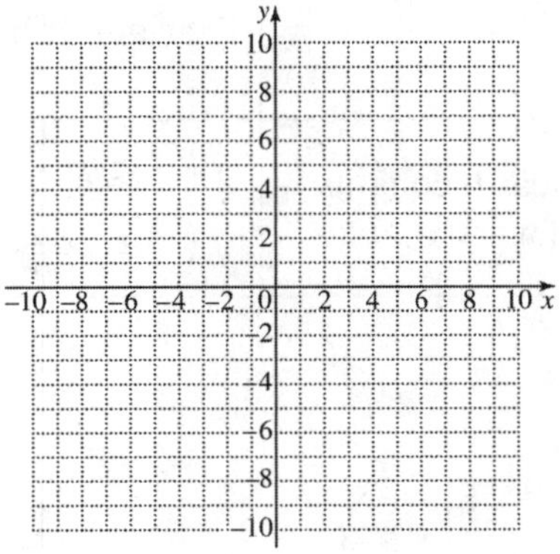

3. $3x - 4y = 6$

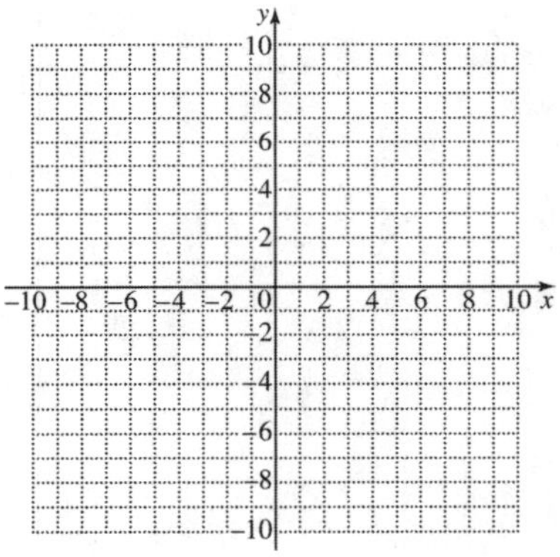

4. Find the slope of a line that passes through $(-1,4)$ and $(3,-6)$.

4. ______________________

Name: **Date:**

Instructor: **Section:**

Graph each line.

5. $f(x) = \frac{2}{3}x - 5$

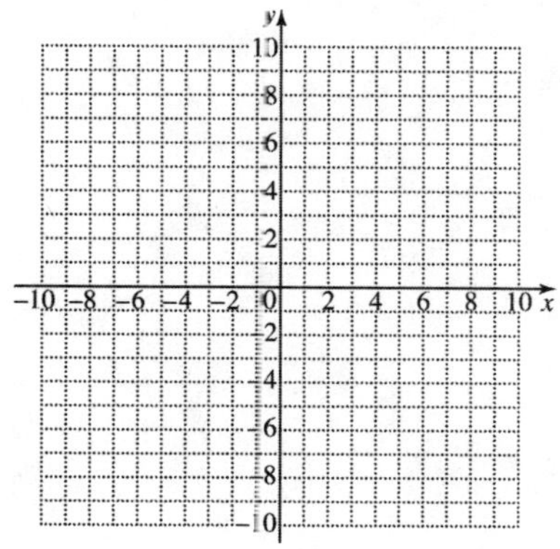

6. $x = -5$

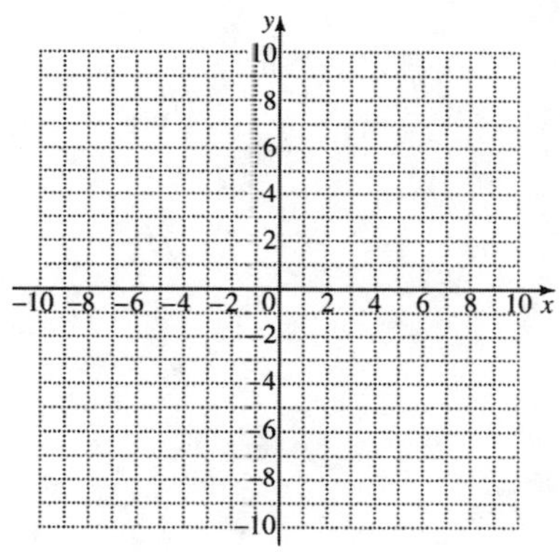

Graph each nonlinear equation.

7. $f(x) = (x+3)^2 + 5$

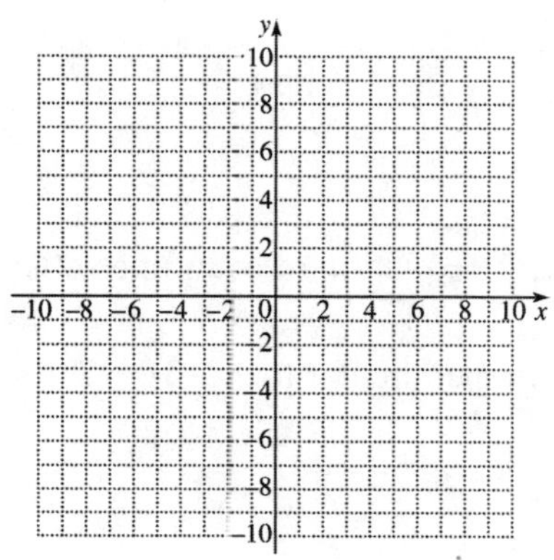

Name: **Date:**

Instructor: **Section:**

8. $f(x) = |x - 3| - 4$

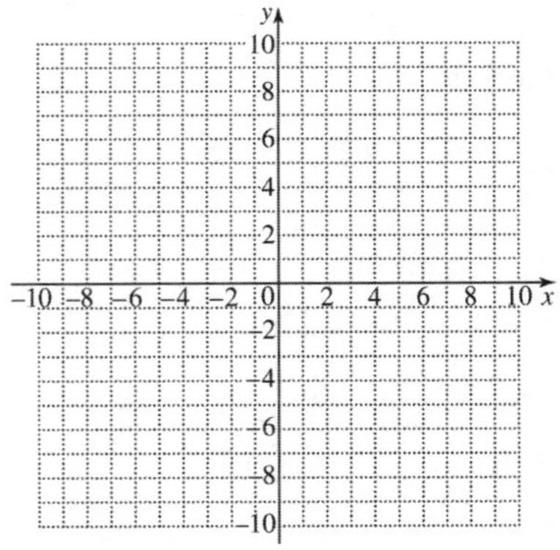

9. $f(x) = -\sqrt{x - 2}$

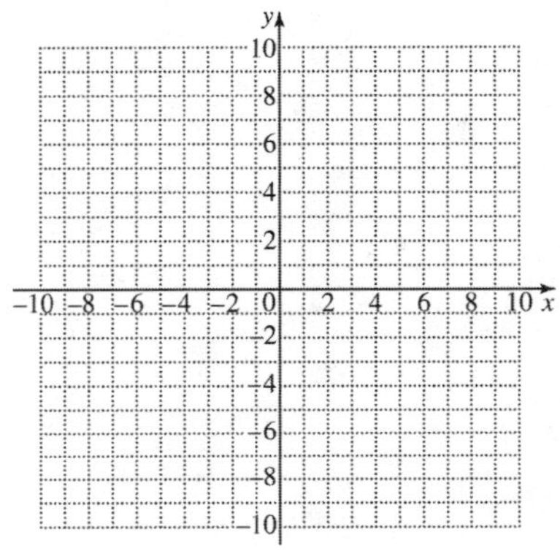

10. Find the slope and y-intercept of the line $8x - 5y = 12$

10. ____________________

Find the equation of each line satisfying the given conditions. Write the equation in standard form.

11. Perpendicular to $x - 3y = 7$ through $(-6, 2)$.

11. ____________________

Name: Date:

Instructor: Section:

12. Vertical line through $\left(-\frac{2}{5}, \frac{1}{3}\right)$

12. ______________________

13. Through $(-2,-4)$ and $(3,8)$.

13. ______________________

14. Parallel to $-x+3y=6$ through $(-1,-4)$.

14. ______________________

Find the domain and range of each relation. Also determine whether the relation is a function.

15.

15. ______________________

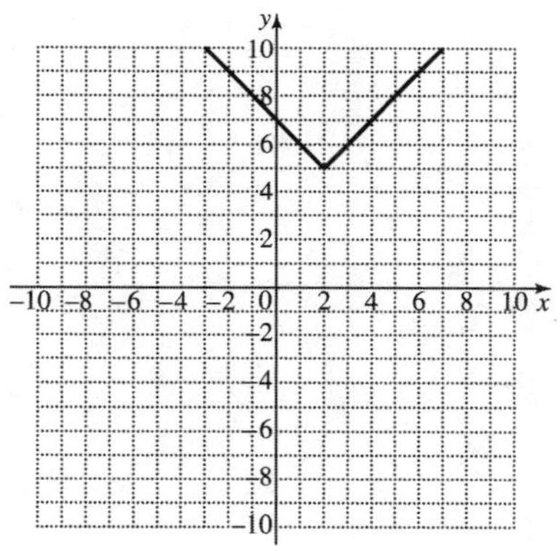

Graph the given inequality.

16. The intersection of $5x-3y<12$ and $y \geq 3$.

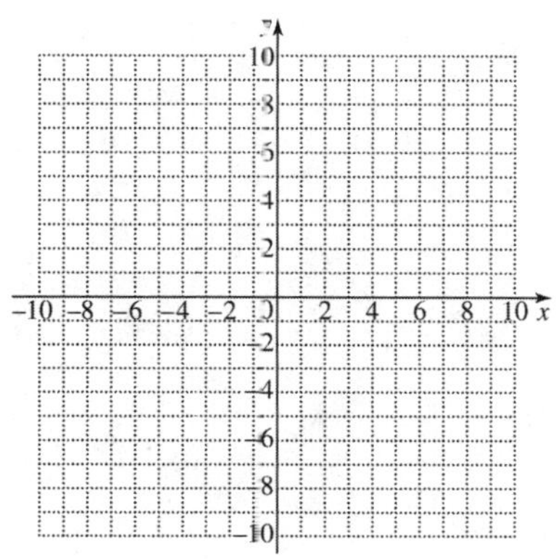

Name: **Date:**

Instructor: **Section:**

17. $-x - y \le 4$

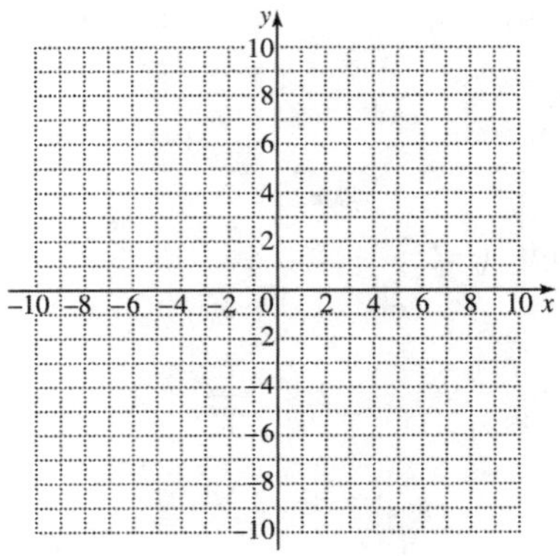

Graph each function.

18. $f(x) = \begin{cases} -2x & \text{if } x > 4 \\ x - 4 & \text{if } x \le -1 \end{cases}$

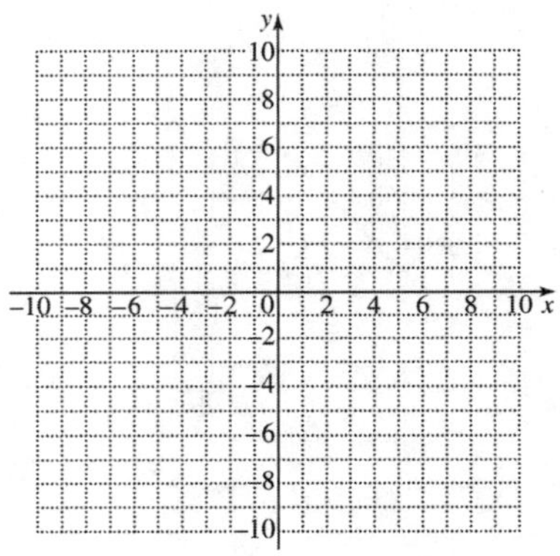

19. $f(x) = \begin{cases} -4 & \text{if } x \ge 3 \\ 2.5x - 5 & \text{if } x < 0 \end{cases}$

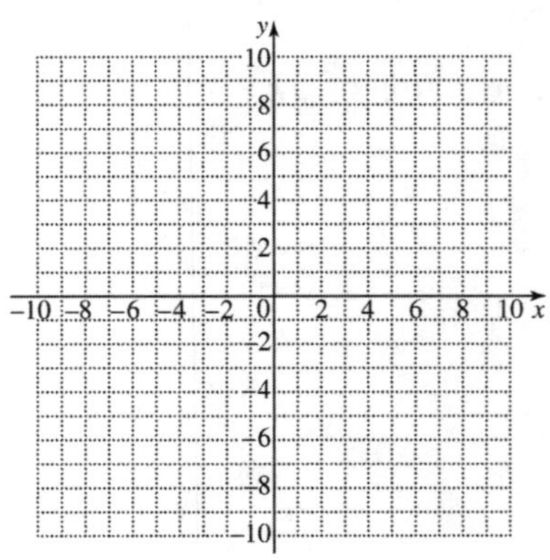

Name: **Date:**

Instructor: **Section:**

Determine if the following lines are parallel, perpendicular, or neither.

20. $2y = 3x + 5$
    $6x - 4y = 8$

20. ____________________

Name: **Date:**

Instructor: **Section:**

Practice Test B

1. What quadrant does the point $(6,-2)$ fall?

1. ____________________

a. I b. II

c. III d. IV

2. Find the slope of the line $15x-18y=20$.

2. ____________________

a. $5/6$ b. $10/9$

c. $-5/6$ d. $-10/9$

3. Find the y-intercept of the line $15x-18y=20$.

3. ____________________

a. $5/6$ b. $10/9$

c. $-5/6$ d. $-10/9$

Find an equation of each line satisfying the given conditions.

4. Horizontal line through the point $(-6,-9)$.

4. ____________________

a. $y=-9$ b. $y=-6$

c. $x=-9$ d. $x=-6$

5. Through the points $(1,7)$ and $(-4,-8)$.

5. ____________________

a. $y=3x-22$ b. $y=3x+4$

c. $y=-3x+24$ d. $y=-3x-4$

Name: **Date:**

Instructor: **Section:**

6. Parallel to $2x+3y=1$ through $(0,-4)$.

6. ____________________

a. $y=-\frac{2}{3}x-4$ b. $y=\frac{3}{2}x-4$

c. $2x+3y=-4$ d. $3x+2y=-4$

Graph the following equations.

7. $-x-y=2$

7. ____________________

a.

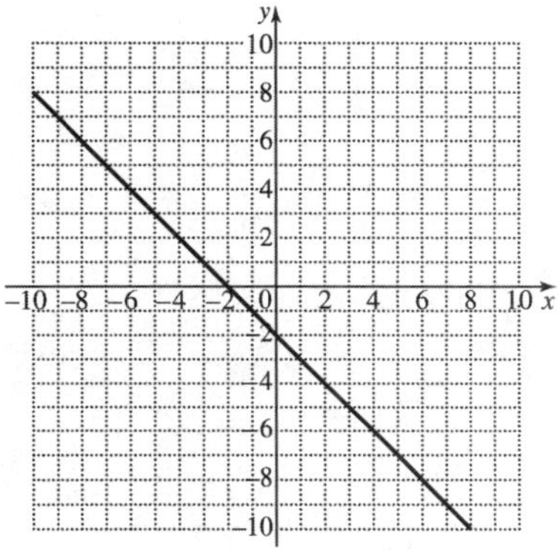

b.

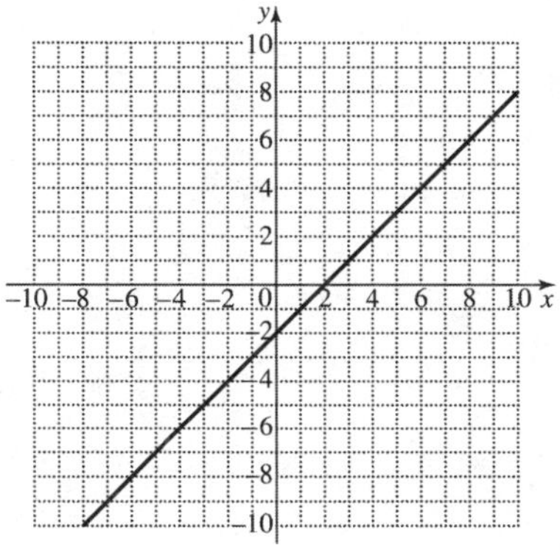

c.

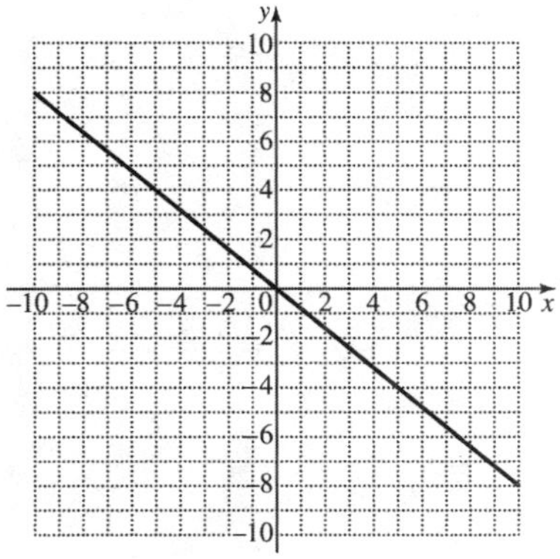

d.

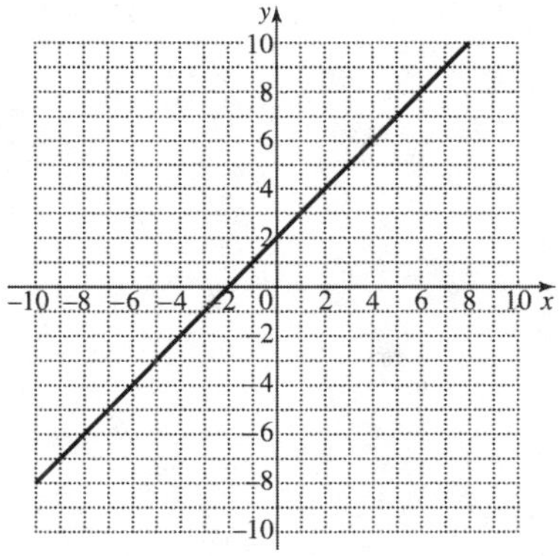

Name: **Date:**

Instructor: **Section:**

8. $f(x) = x^2 - 6$

8. ____________________

a.

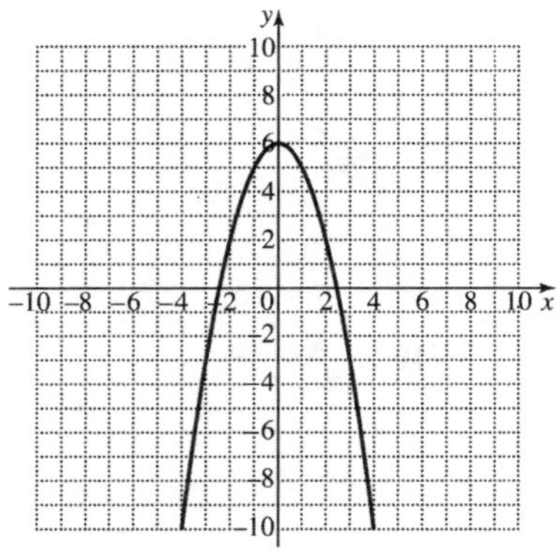

b.

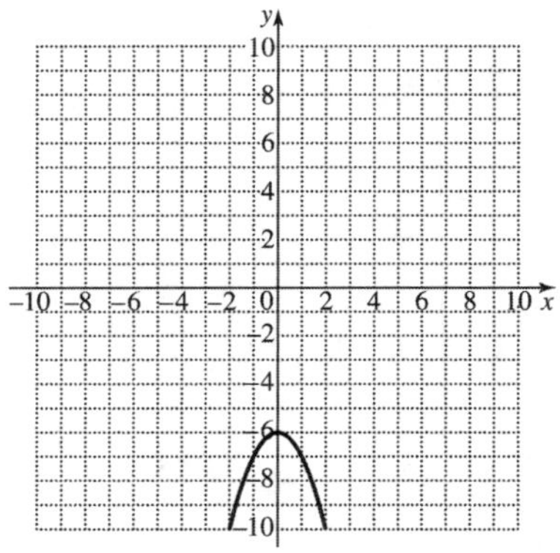

c.

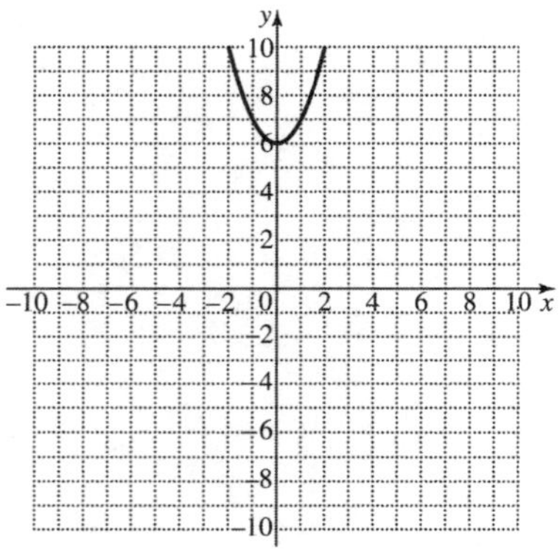

d.

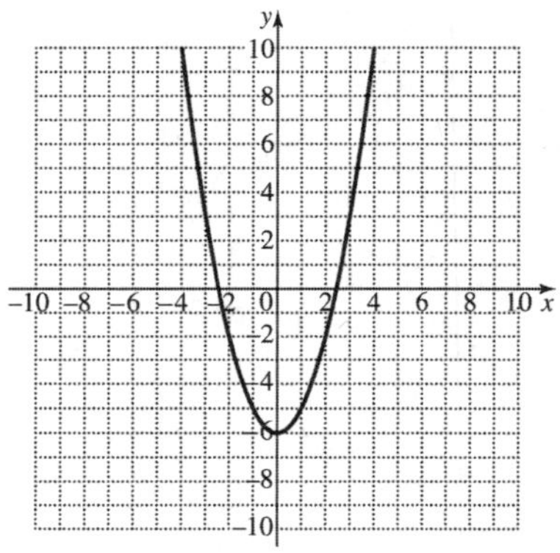

9. $f(x) = |x + 3|$

9. ____________________

a.

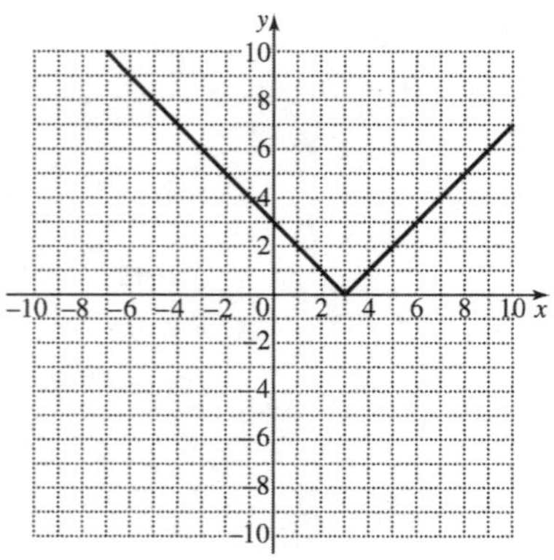

b.

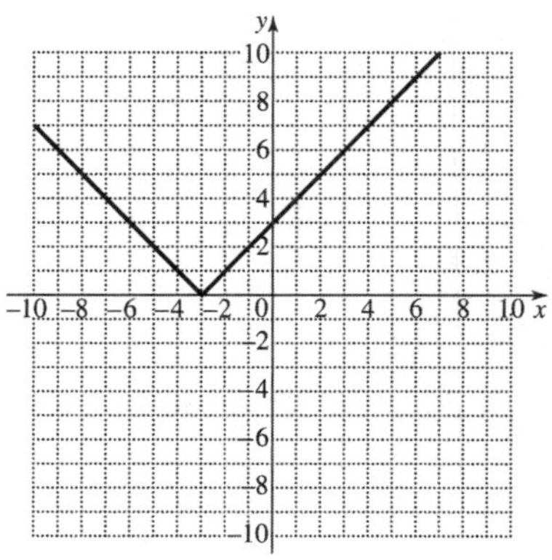

Name: **Date:**

Instructor: **Section:**

c.

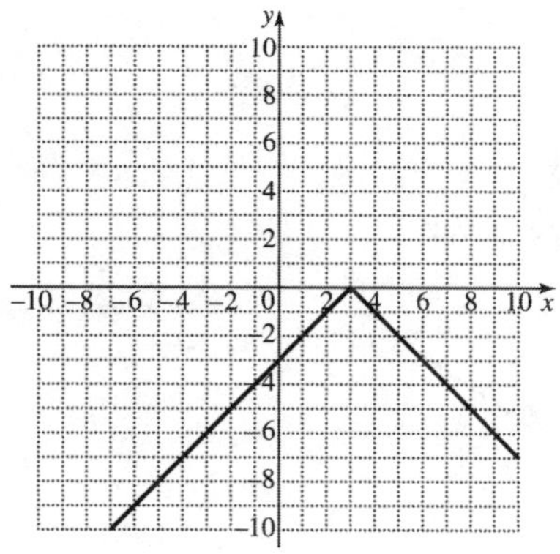

d.

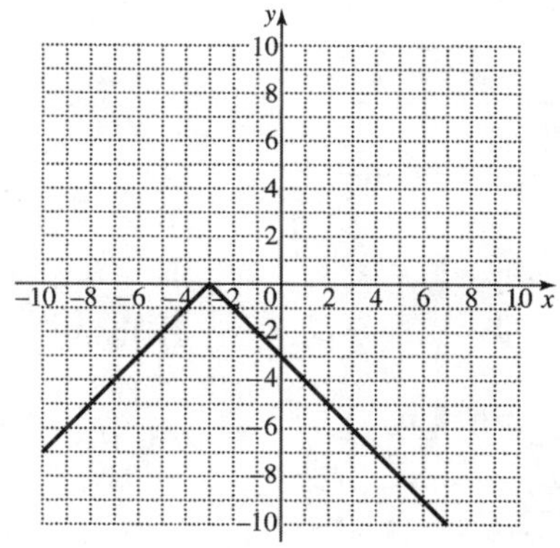

10. $f(x) = -(x+1)^2 - 3$

10. ______________________

a.

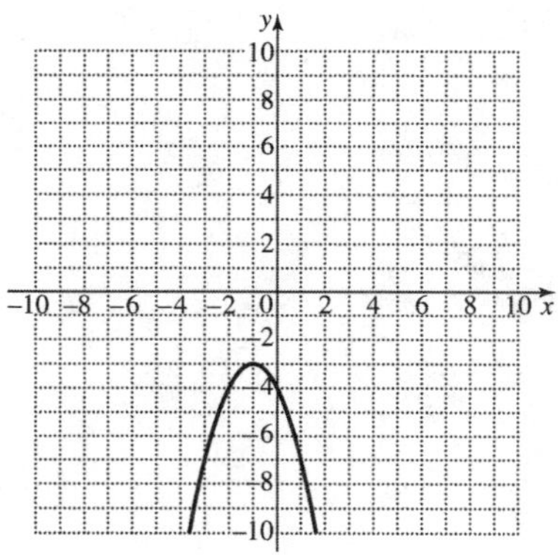

b.

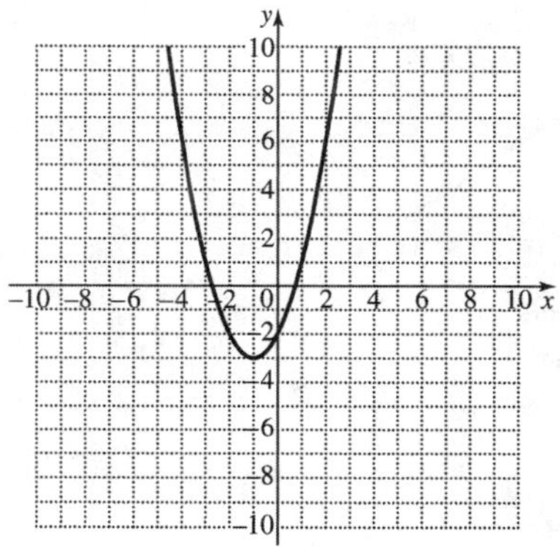

c.

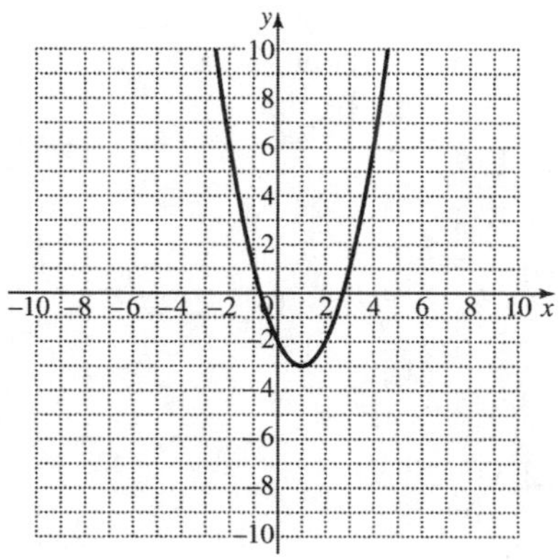

d.

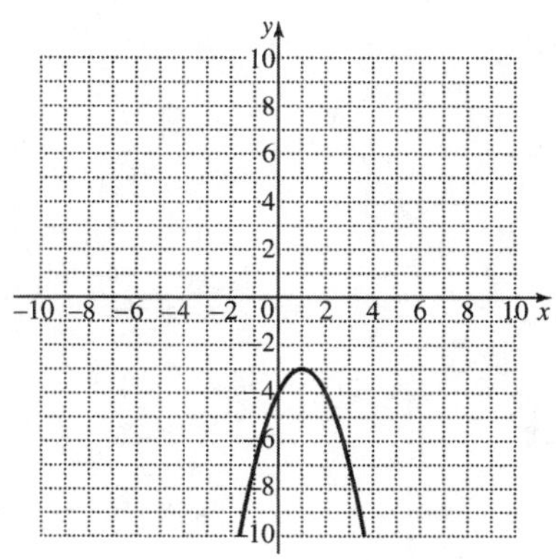

Name: **Date:**

Instructor: **Section:**

11. $y = -\frac{2}{3}x - 6$

11. ______________________

a.

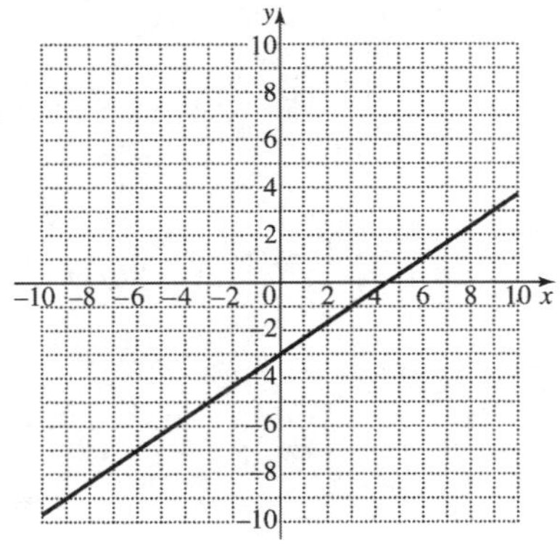

b.

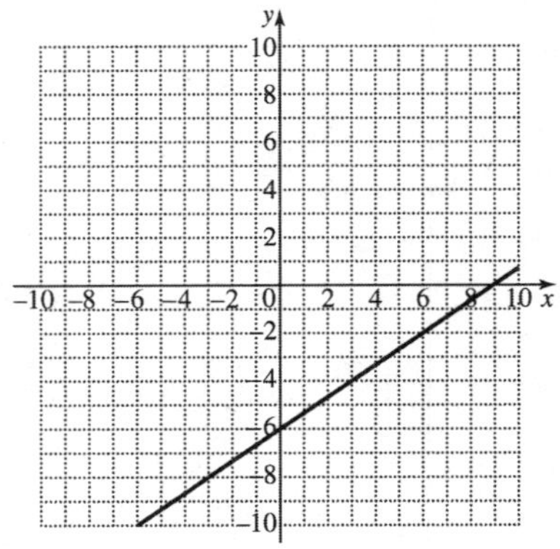

c.

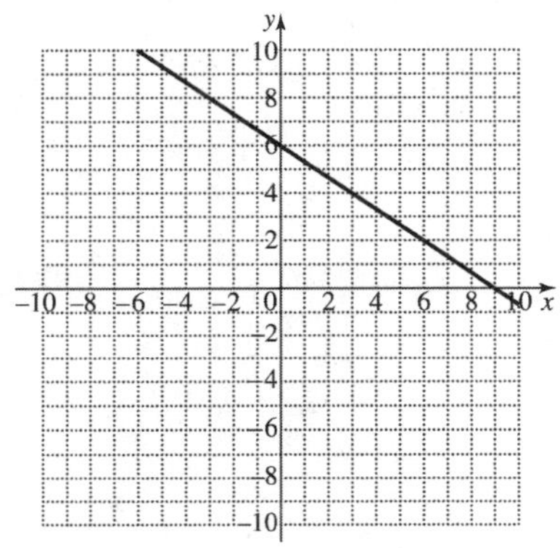

d.

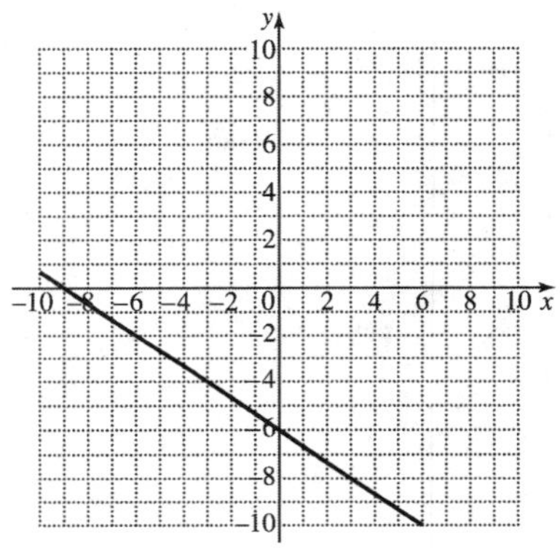

12. $5x - 4y = 12$

12. ______________________

a.

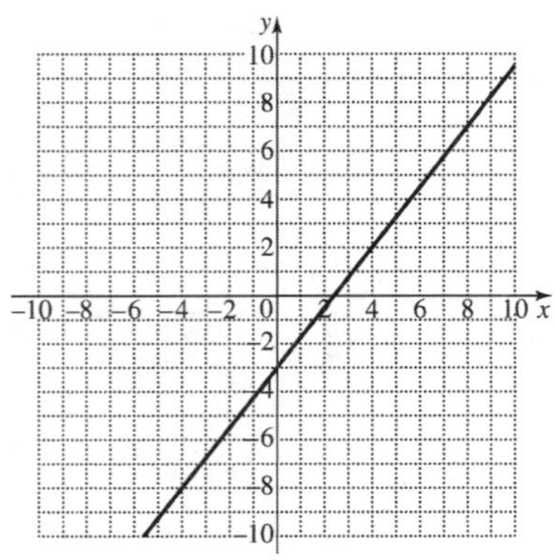

b.

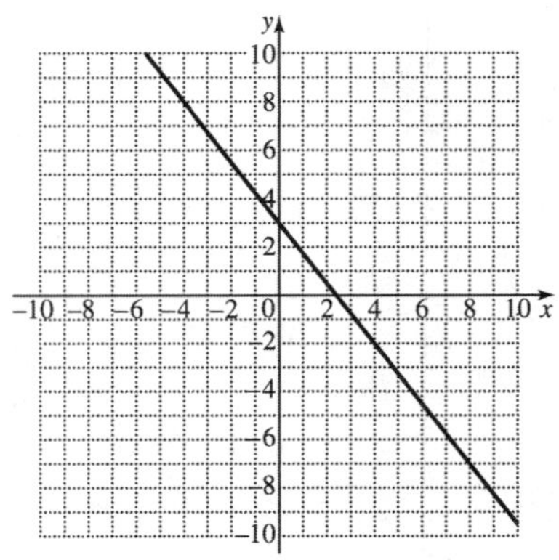

c.

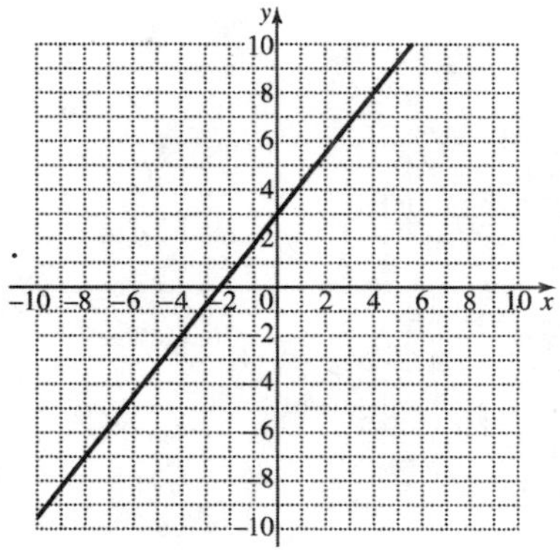

d.

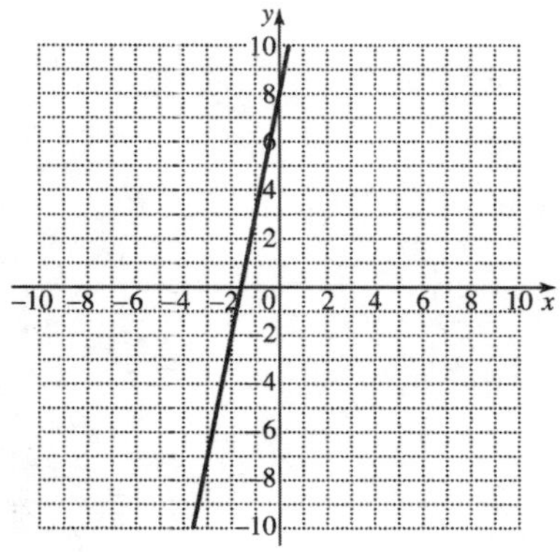

Graph each inequality.

13. $y < 5$

13. ____________________

a.

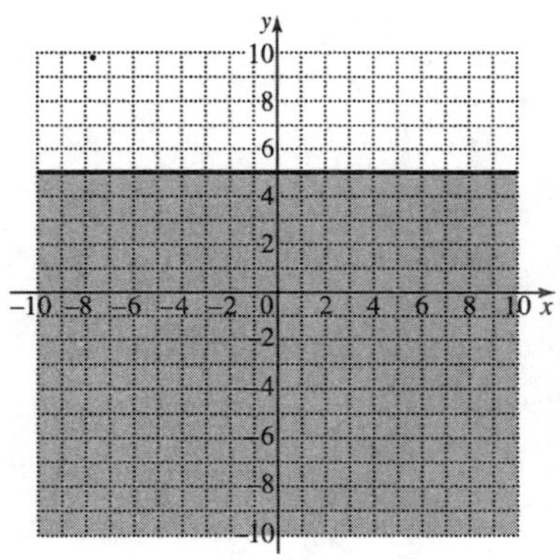

b.

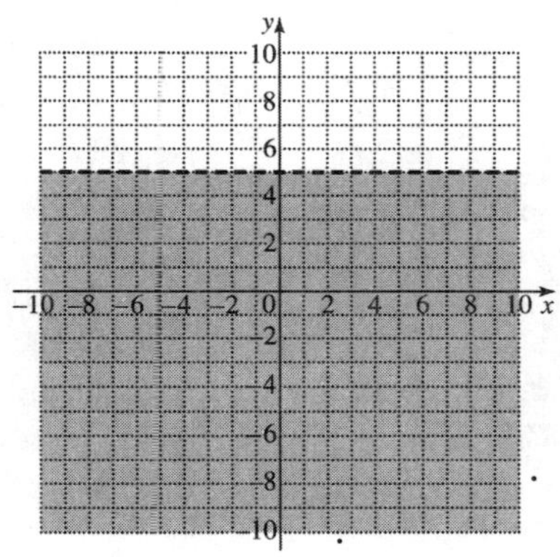

c.

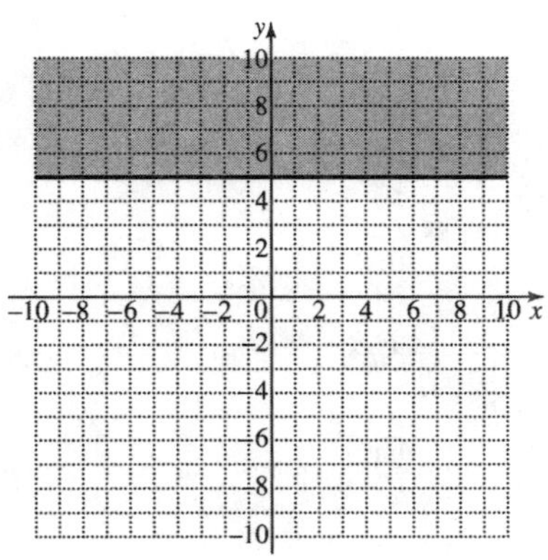

d.

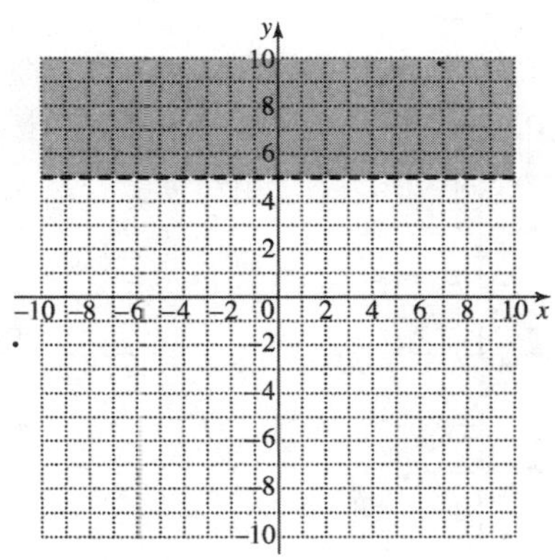

Name: **Date:**

Instructor: **Section:**

14. $y < 2x - 6$

14. ____________________

a.

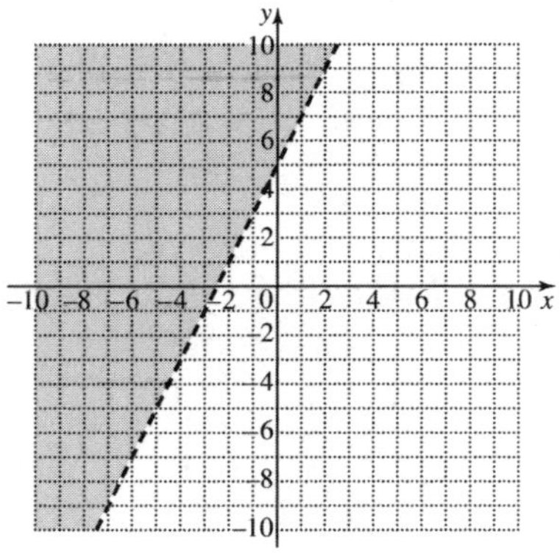

b.

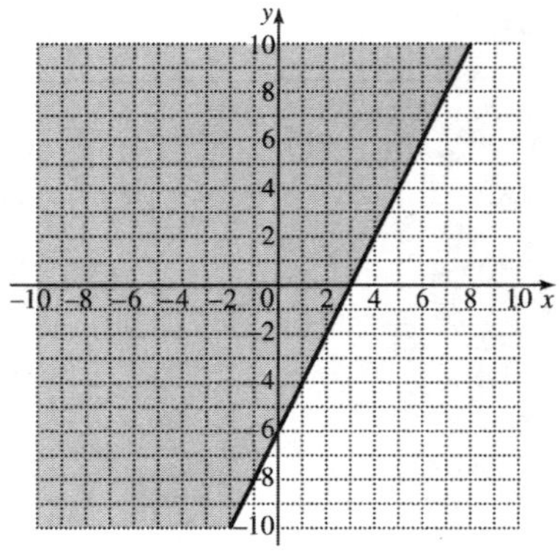

c.

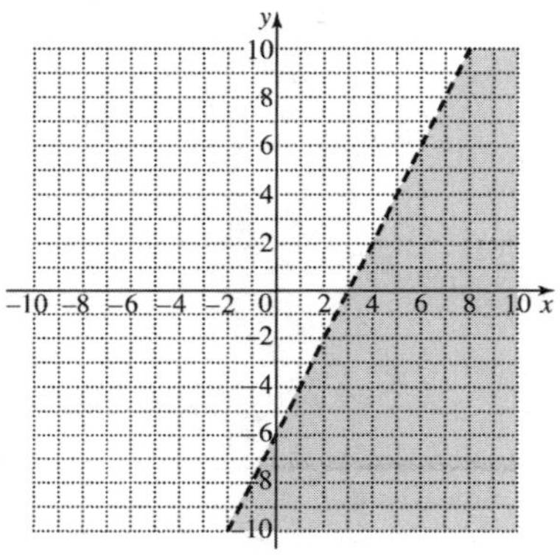

d.

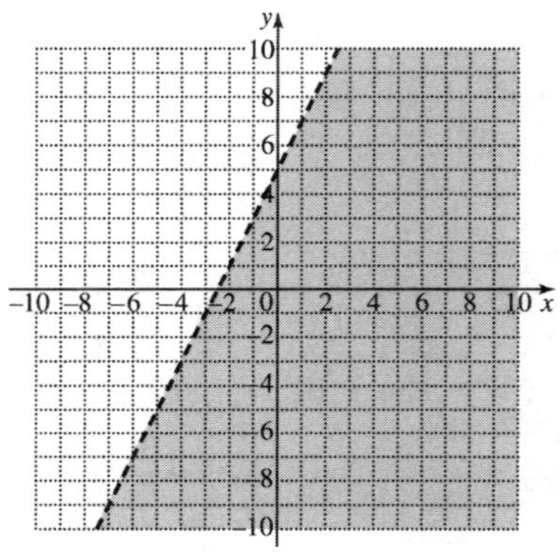

15. The intersection of $x + y > 3$ and $x \leq 5$.

15. ____________________

a.

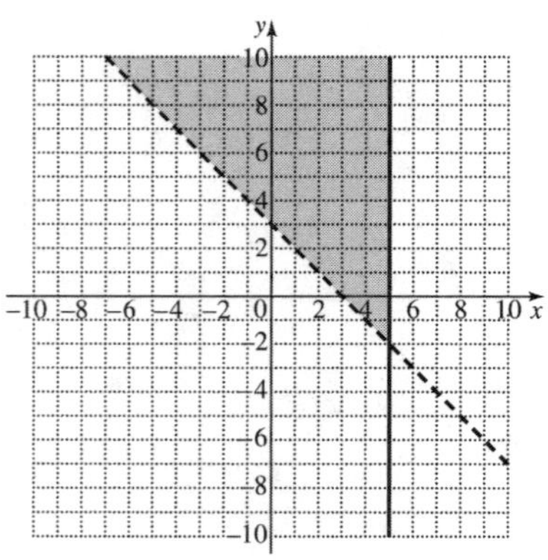

b.

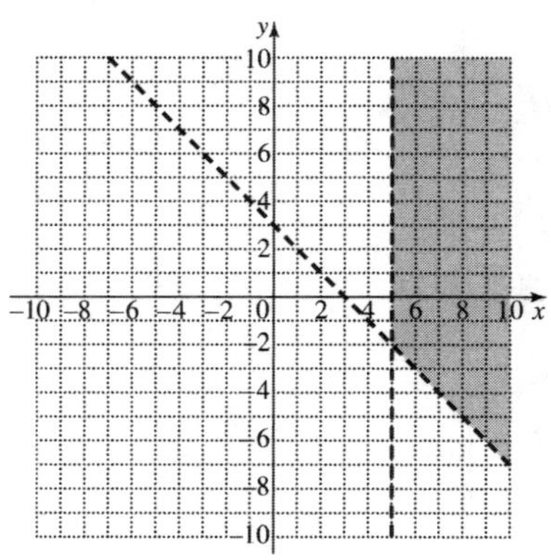

Name: **Date:**

Instructor: **Section:**

c.

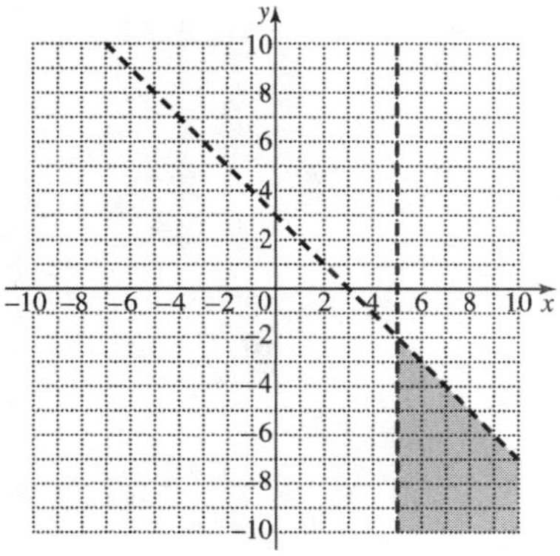

d.

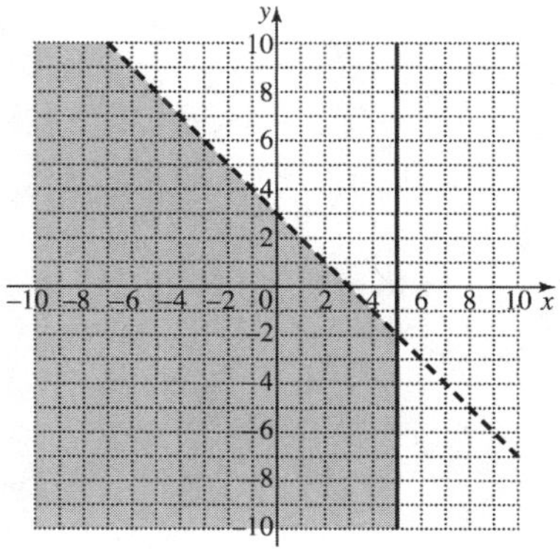

Find the domain and range of the following function.

16. 16. ____________________

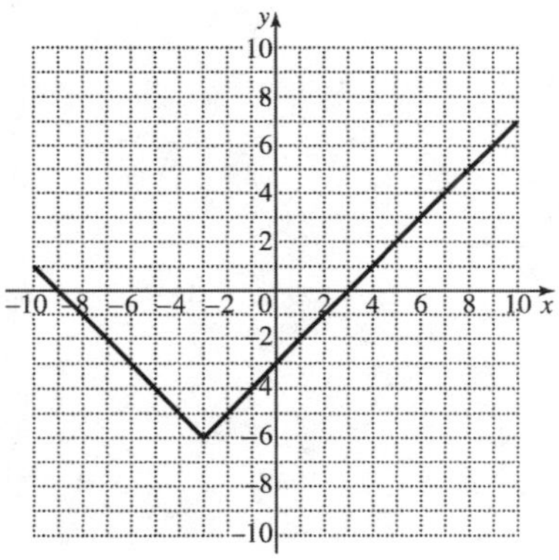

a. $D=[-3,\infty)$ $R=[-6,\infty)$

b. $D=(-3,\infty)$ $R=(-6,\infty)$

c. $D=[-3,\infty)$ $R=(-\infty,\infty)$

d. $D=(-\infty,\infty)$ $R=[-6,\infty)$

Name: **Date:**

Instructor: **Section:**

17. Determine whether the following is parallel, perpendicular, or neither.

$$6x+5y=18$$
$$5x-6y=24$$

17. ____________________

a. Parallel

b. Perpendicular

c. Neither

Graph each function.

18. $f(x)=\begin{cases}-x-3 & if\ \ x>3\\ 2x+1 & if\ \ x\leq -1\end{cases}$

18. ____________________

a.

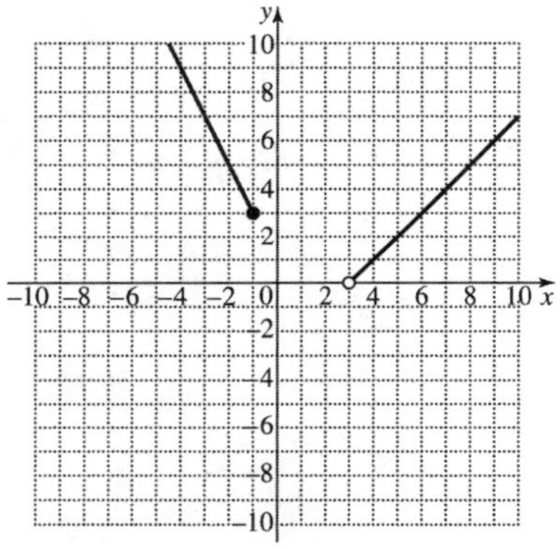

b.

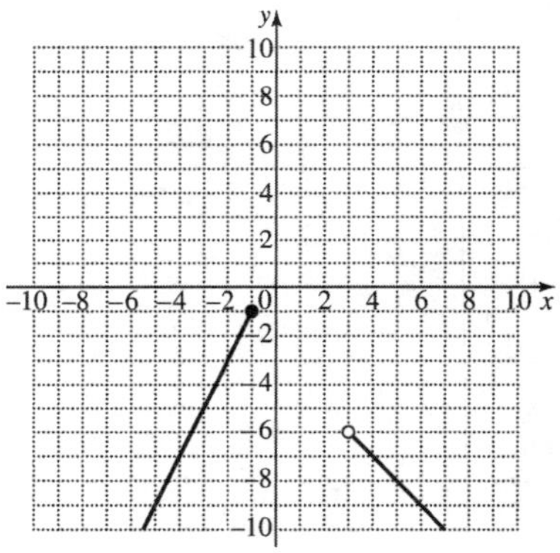

c.

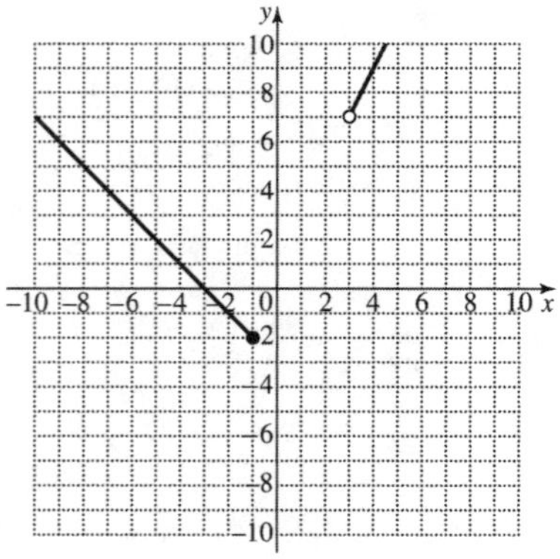

d.

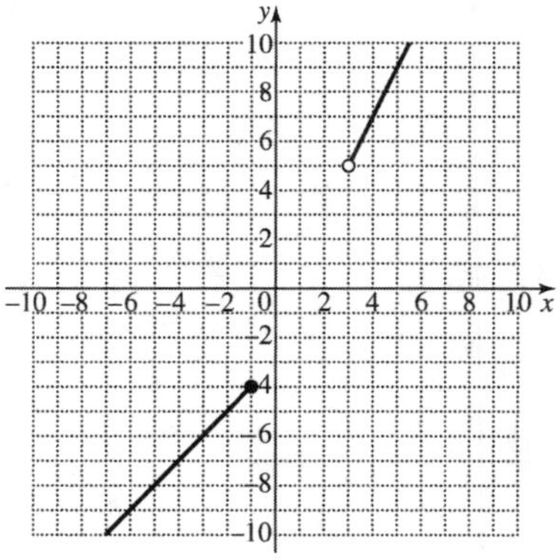

Name: Date:

Instructor: Section:

19. $f(x)=\begin{cases} x & \text{if } x>0 \\ -x & \text{if } x\le 0 \end{cases}$

19. ____________________

a.

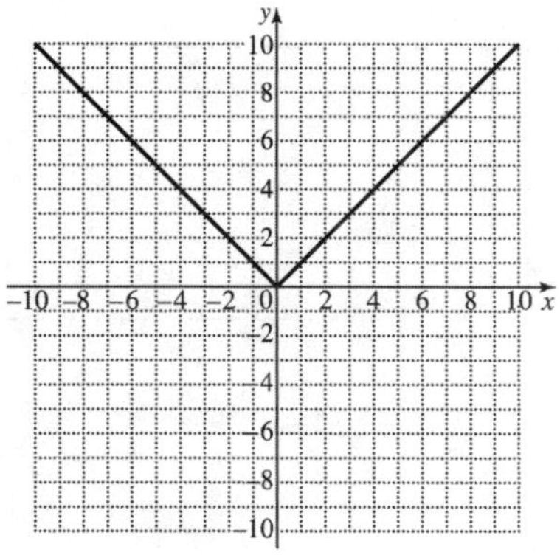

b.

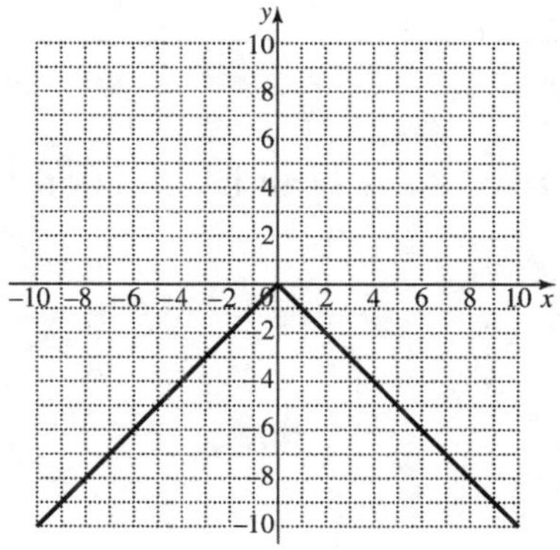

c.

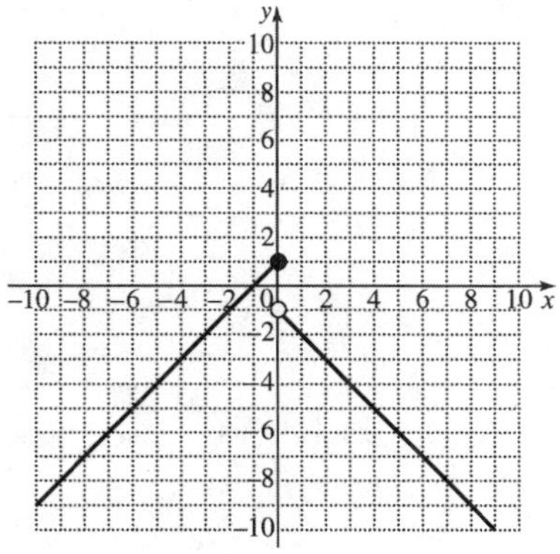

d.

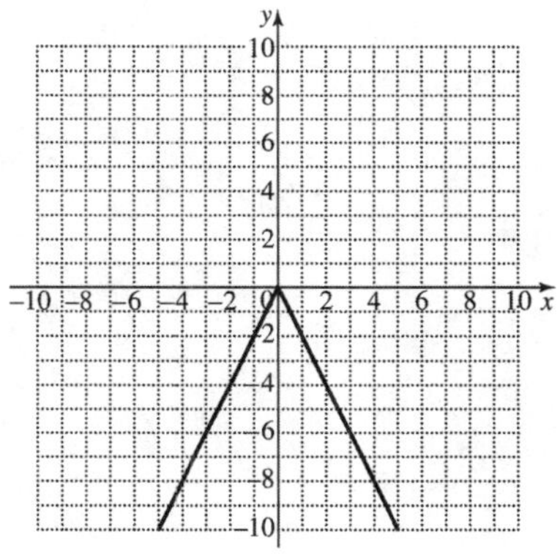

20. Determine whether $y=|x|+3$ is linear.

20. ____________________

a. Yes it is linear

b. No, it is not linear.

Name: **Date:**

Instructor: **Section:**

Chapter 4 Systems of Equations
Section 4.1 Solving Systems of Linear Equations in Two Variables

Learning Objectives

1. Determine whether an ordered pair is a solution of a system of two linear equations.
2. Solve a system by graphing.
3. Solve a system by substitution.
4. Solve a system by elimination.

Objective 1

Determine whether each given ordered pair is a solution of each system.

1. $(3,5)$

$$\begin{cases} 2x-3y=-9 \\ 4x+2y=-2 \end{cases}$$

1. ______________________

2. $(7,0)$

$$\begin{cases} x+3y=7 \\ 2x+5y=14 \end{cases}$$

2. ______________________

3, $(-2,-5)$

$$\begin{cases} 4x-y=3 \\ y=\frac{1}{2}x+4 \end{cases}$$

3. ______________________

4. $(-4,8)$

$$\begin{cases} y=-2x \\ x=\frac{1}{4}y-6 \end{cases}$$

4. ______________________

Name: **Date:**

Instructor: **Section:**

Objective 2

Solve each system of equations by graphing.

5. $\begin{cases} x + y = 6 \\ 2x + y = 4 \end{cases}$

5. ______________________

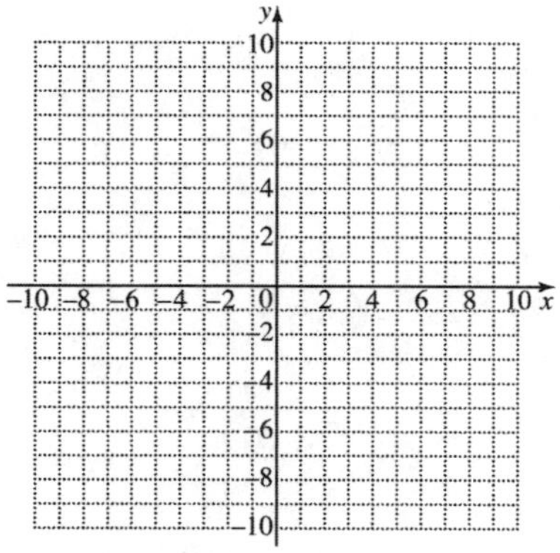

6. $\begin{cases} 2x - 3y = -3 \\ x - 4y = 6 \end{cases}$

6. ______________________

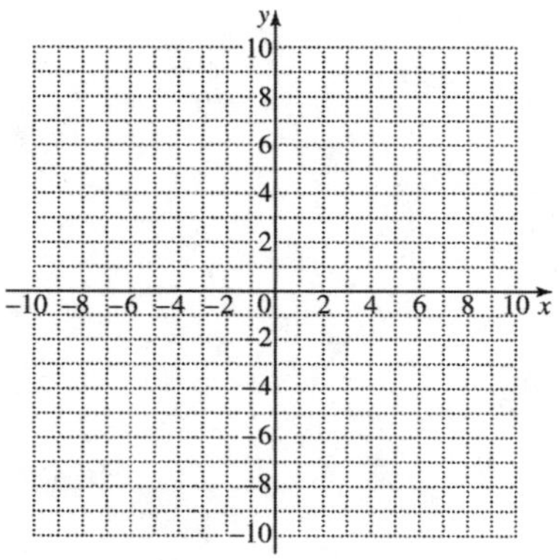

Name: **Date:**

Instructor: **Section:**

7. $\begin{cases} y = \frac{2}{3}x - 10 \\ y = -\frac{2}{5}x + 6 \end{cases}$

7. ____________________

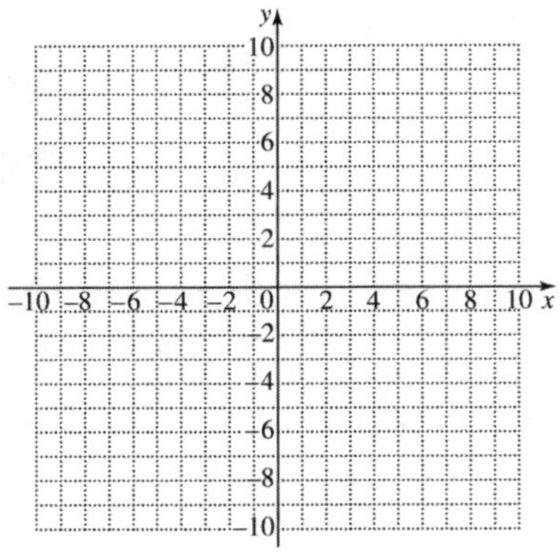

Objective 3

Solve each system of equation by the substitution method.

8. $\begin{cases} 4x - y = 9 \\ 2x + 3y = -27 \end{cases}$

8. ____________________

9. $\begin{cases} x + y = 7 \\ 2x - 4y = 2 \end{cases}$

9. ____________________

Name: **Date:**

Instructor: **Section:**

10. $\begin{cases} \frac{1}{4}x - \frac{1}{3}y = \frac{2}{3} \\ \frac{1}{4}x + \frac{1}{5}y = \frac{2}{5} \end{cases}$

10. ____________________

11. $\begin{cases} x = 3y + 2 \\ 5x - 15y = 10 \end{cases}$

11. ____________________

Objective 4

Solve the system of equations by the elimination method.

12. $\begin{cases} 4x - 3y = 7 \\ 2x + y = 11 \end{cases}$

12. ____________________

13. $\begin{cases} 8x + 6y = 10 \\ 3y = -4x + 5 \end{cases}$

13. ____________________

Name: **Date:**

Instructor: **Section:**

14. $\begin{cases} 3x = 2y - 4 \\ 4x + 3y = 6 \end{cases}$

14. ____________________

15. $\begin{cases} 10x - 10y = -40 \\ 5x + 5y = 20 \end{cases}$

15. ____________________

Concept Extension

Solve this system of equations.

16. $\begin{cases} x = y + 2 \\ 3x + 2y = 6 \\ x - 2y = 2 \end{cases}$

16. ____________________

Name: **Date:**

Instructor: **Section:**

Section 4.2 Solving Systems of Linear Equations in Three Variables

Learning Objectives

1. Solve a system of three linear equations in three variables.

Objective 1

Solve each system.

1. $\begin{cases} x+y-z=4 \\ 3x+2y-3z=9 \\ x-y+4z=8 \end{cases}$

1. ______________________

2. $\begin{cases} x+2y=4 \\ x+z=-3 \\ 3x+y-z=5 \end{cases}$

2. ______________________

3. $\begin{cases} x+2y-z=4 \\ 3x-y+2z=6 \\ x+y=5 \end{cases}$

3. ______________________

Name: **Date:**

Instructor: **Section:**

4. $\begin{cases} x - y + 2z = 10 \\ 2x - 4y + z = -2 \\ x + 3y - z = 6 \end{cases}$

4. ______________________

5. $\begin{cases} 2x - 3y + z = 2 \\ x - 5y + 5z = 3 \\ 3x + y - 3z = 5 \end{cases}$

5. ______________________

6. $\begin{cases} x + 3y - 4z = 6 \\ 3x + 9y - 12x = 18 \\ -5x - 15y + 20z = -30 \end{cases}$

6. ______________________

Name: **Date:**

Instructor: **Section:**

7. $\begin{cases} x - z = 0 \\ 2x + y + z = 7 \\ 3y - 2z = 10 \end{cases}$

7. ____________________

8. $\begin{cases} 8x - 2y + 6z = 28 \\ 3x + y - 2z = 3 \\ 2x - 6y + 3z = -5 \end{cases}$

8. ____________________

9. $\begin{cases} 4x + 4y - 3z = -22 \\ 5x - 6y - 4z = 15 \\ 3x + 5y + z = -5 \end{cases}$

9. ____________________

Name: **Date:**

Instructor: **Section:**

10. $\begin{cases} 2x+4y+3z=-9 \\ 3x+2y+5z=-19 \\ 4x-5y+2z=-19 \end{cases}$

Concept Extension

11.

$$\begin{cases} 2x+4y+2z+3w=-7 \\ -3x-2y+5z+4w=9 \\ 4x-3y-2z-3w=11 \\ -2x+3y-5z-4w=-16 \end{cases}$$

11. ____________________

Name: **Date:**

Instructor: **Section:**

Section 4.3 Systems of Linear Equations and Problem Solving

Learning Objectives

1. Solving problems that can be modeled by a system of two linear equations.
2. Solve problems with cost and revenue functions.
3. Solve problems that can be modeled by a system of three linear equations.

Objective 1

Solve.

1. One number is five more than twice the second number. If the difference between twice the first number and three times the second number is six, what are the two numbers?

1. ____________________

2. Find how many quarts of 4% butterfat milk and 1% butterfat milk should be mixed to yield 60 quarts of 2% butter fat milk.

2. ____________________

3. Lucy went to the local pawn shop. She bought \$18 bracelets and \$25 necklaces. If Lucy bought a total of 26 pieces and spent only \$566, how many of each type did she purchase?

3. ____________________

Name: **Date:**

Instructor: **Section:**

4. The perimeter of a rectangle is 52 feet. The length of the rectangle is 6 more than triple the width. Find the dimensions of the rectangle.

4. ______________________

5. Two joggers leave an apartment building and jog in opposite directions. One jogger travels at 2 mph faster than the other jogger. If after 2.5 hours, they are 15 miles apart, what is the rate of each jogger?

5. ______________________

Objective 2

Given the cost function and the revenue function, find the number of units x that must be sold to break even.

6. $C(x) = 1.7x + 1700$ $R(x) = 2.4x$

6. ______________________

7. $C(x) = 19x + 2250$ $R(x) = 34x$

7. ______________________

Name: **Date:**

Instructor: **Section:**

8. The planning department of Abstract Office Supplies has been asked to determine whether the company should introduce a new computer desk next year. The department estimates that $6000 of new manufacturing equipment will need to be purchased and that the cost of constructing each desk will be $200. The department also estimates that the revenue from each desk will be $450.

   a. Determine the revenue function $R(x)$ from the sale of x desks.

   8a. ____________________

   b. Determine the cost function $C(x)$ for manufacturing x desks.

   8b. ____________________

   c. Find the break-even point.

   8c. ____________________

9. The U-Haul-It company is looking to purchase a new $20,000 truck for its fleet. The company rents out the trucks for $100 each rental. When the truck is returned, it will cost U-Haul-it $50 to clean and get the truck ready for its next renter. When will the company finally start earning money for this vehicle.

   9. ____________________

Name: **Date:**

Instructor: **Section:**

Objective 3

10. One number is four less than the second number. Triple the first number is 1 more than the third number. The sum of all three numbers is 23. Find the numbers.

10. ____________________

11. The sum of the angles of a triangle is 180 degrees. The sum of the twice the smallest angle and the largest angle is 175 degrees. Three times the second angle less the largest angle is 65 degrees. Find the measures of the angles of the triangle.

11. ____________________

Concept Extension

12. Find the values of a, b, and c such that the equation $y = ax^2 + bx + c$ has the ordered pair solutions of $(3,27)$, $(-5,131)$, and $(-2,32)$.

12. ____________________

Name: **Date:**

Instructor: **Section:**

Section 4.4 Solving Systems of Equations by Matrices

Learning Objectives

1. Use matrices to solve a system of two equations.
2. Use matrices to solve system of three equations.

Vocabulary

Use the choices to complete each statement.

Column(s)	**Element**
Matrix	**Row(s)**

1. Any two ______________ can be interchanged.
2. A rectangular array of number is called a ________________.
3. A(n)________________ is what we call each number in the matrix.
4. The numbers arranged vertically in a line are in the same ___________________.

Objective 1

Solve each system of linear equations using matrices.

5. $\begin{cases} x+3y=2 \\ x+2y=0 \end{cases}$ 5. ______________________

6. $\begin{cases} 2x+3y=12 \\ x-2y=6 \end{cases}$ 6. ______________________

Name: **Date:**

Instructor: **Section:**

7. $\begin{cases} y = -6x + 7 \\ 4x - y = 13 \end{cases}$

7. ______________________

8. $\begin{cases} 9x + 8y = 36 \\ 3x + 5y = 5 \end{cases}$

8. ______________________

9. $\begin{cases} 2x - 5y = 8 \\ x = 4 + \frac{5}{2}y \end{cases}$

9. ______________________

Name: **Date:**

Instructor: **Section:**

10. $\begin{cases} 4x + 7y = 19 \\ 8x + 14y = -38 \end{cases}$

10. ______________________

11. $\begin{cases} 5x + y + 3z = 1 \\ 2x - 2y + 6z = -14 \\ x - y = -1 \end{cases}$

11. ______________________

Name: **Date:**

Instructor: **Section:**

12. $\begin{cases} 4x - 3y + 12z = -108 \\ 6x - 4y - 3z = -72 \\ 2x - y - 2z = -16 \end{cases}$

12. ____________________

Concept Extension

Solve the given system of equations by using matrices.

13. $\begin{cases} 2x + 2y + 2z + 2w = 10 \\ 2x + y + z + w = 6 \\ 3x + 3y + 3z = 6 \\ -4x - 4y = 0 \end{cases}$

13. ____________________

Name:

Instructor:

Date:

Section:

Section 4.5 Systems of Linear Inequalities

Learning Objectives

1. Graph a system of linear inequalities.

Objective 1

Graph the solutions of each system of linear inequalities.

1. $\begin{cases} y < -2x + 2 \\ y > x + 4 \end{cases}$

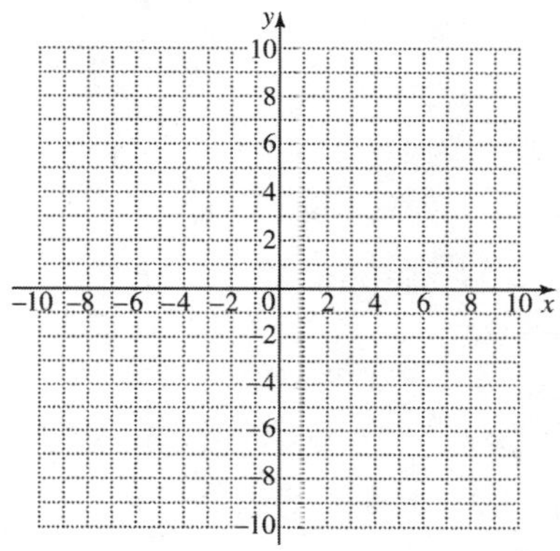

2. $\begin{cases} y \leq x + 6 \\ y > 2x - 8 \end{cases}$

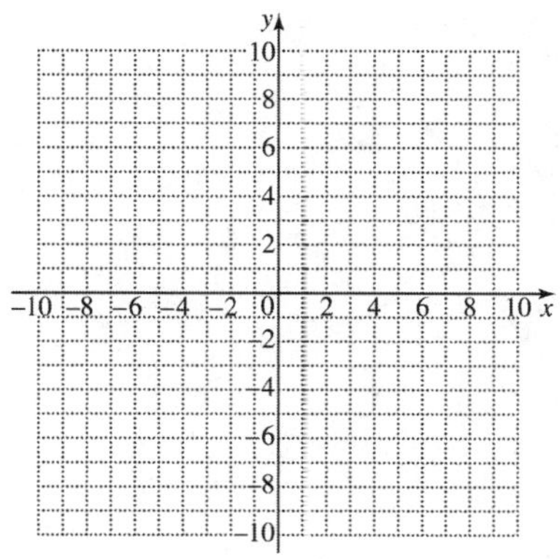

Name: **Date:**

Instructor: **Section:**

3. $\begin{cases} y > -3x - 8 \\ y \le 2x + 4 \end{cases}$

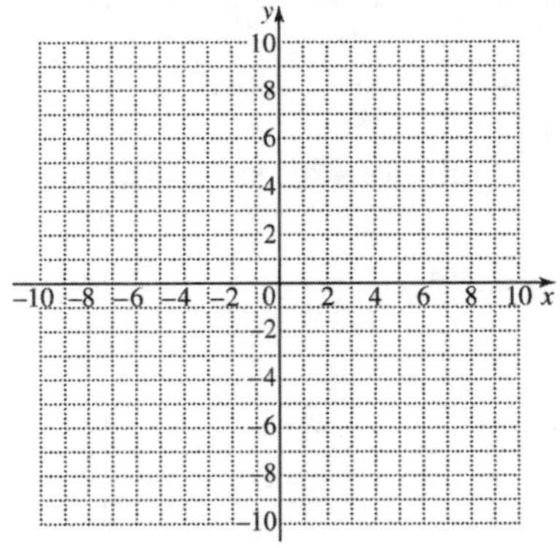

4. $\begin{cases} y \le 3x - 6 \\ x > 3 \end{cases}$

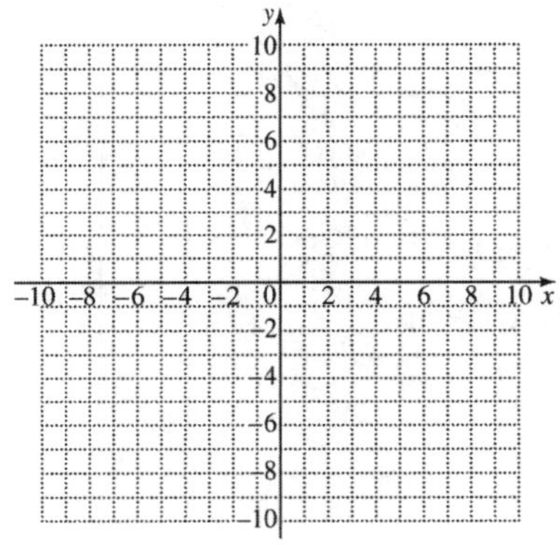

5. $\begin{cases} 3x - 4y < 9 \\ y - 2x \ge -5 \\ x < 2 \end{cases}$

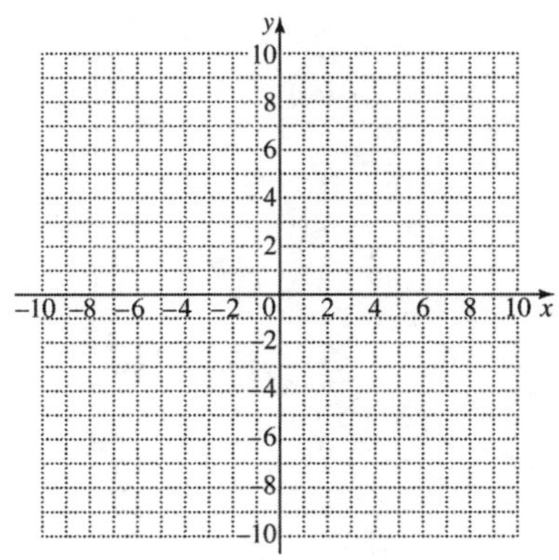

Name: **Date:**

Instructor: **Section:**

6. $\begin{cases} y + 2x \geq 0 \\ 5x - 3y \leq 12 \\ y \leq 2 \end{cases}$

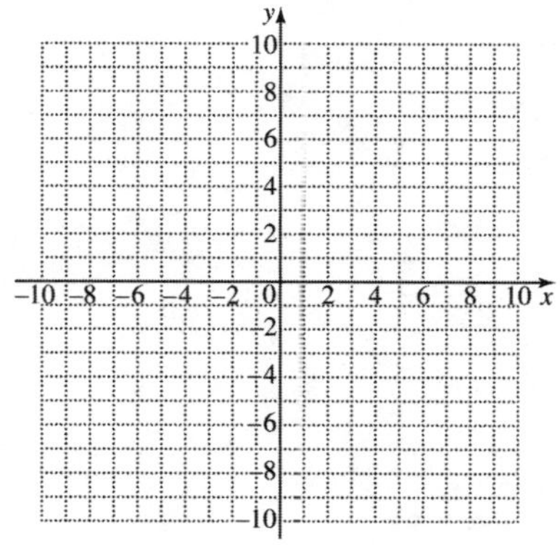

7. $\begin{cases} 3x + 4y < 12 \\ 2x - y > 8 \\ y < -3 \end{cases}$

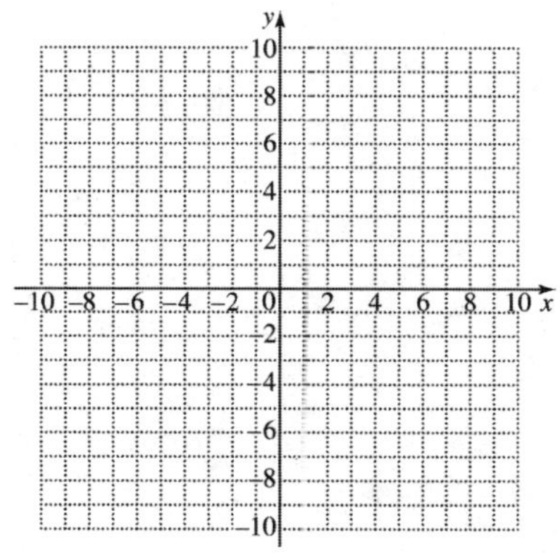

Name: **Date:**

Instructor: **Section:**

Concept Extension

Graph the solutions of the system of linear inequalities.

8. $\begin{cases} 4x + 2y > 6 \\ 3x - 4y > 9 \\ x \leq 5 \\ y \leq 2 \end{cases}$

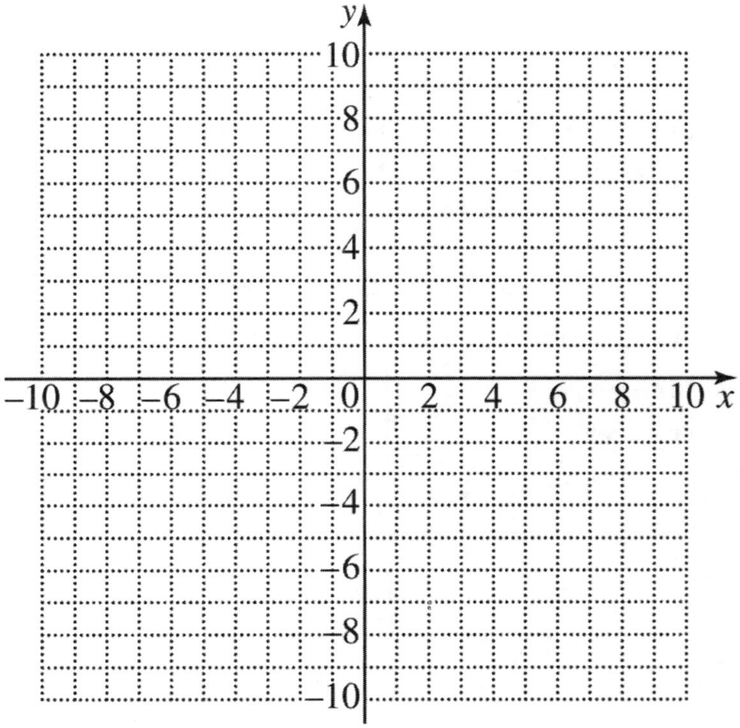

Name: **Date:**

Instructor: **Section:**

Chapter 4 Vocabulary

Vocabulary Word	Definition	Example
System of Linear Equations	A set of two or more equations.	$\begin{cases} x+y=7 \\ 2x+5y=13 \end{cases}$
Solution of a System	An ordered pair (or triple) that satisfies all equations of that system.	$\begin{cases} x+y=7 \\ x-y=-1 \end{cases}$ $(3,4)$ is a solution of the given system.
Consistent System	A system that has at least one solution.	$\begin{cases} x+y=5 \\ x-y=6 \end{cases}$ is consistent
Inconsistent System	A system that does not have a solution.	$\begin{cases} x+2y=6 \\ 4x+8x=10 \end{cases}$ is inconsistent
Dependent System	A system that all equations of the system are identical lines.	$\begin{cases} x+y=2 \\ 4x+4y=8 \end{cases}$ is dependent
Independent System	A system where each equation will result in a distinct line.	$\begin{cases} 4x+5y=7 \\ 2x+3y=4 \end{cases}$ is independent
Matrix	A rectangular array of numbers	$\begin{bmatrix} 4 & 5 \\ 3 & 2 \end{bmatrix}$
Row Operations	Operations done to solve a matrix: 1. Switch rows. 2. Multiply a row by a nonzero constant. 3. Add rows together.	
System of Linear Inequalities	A set of linear inequalities.	$\begin{cases} x+y<4 \\ 2x+6y\geq 7 \end{cases}$

Name: **Date:**

Instructor: **Section:**

Practice Test A

Solve the following system of equations by graphing.

1. $\begin{cases} x + y = 5 \\ x + 2y = 4 \end{cases}$

1. ______________________

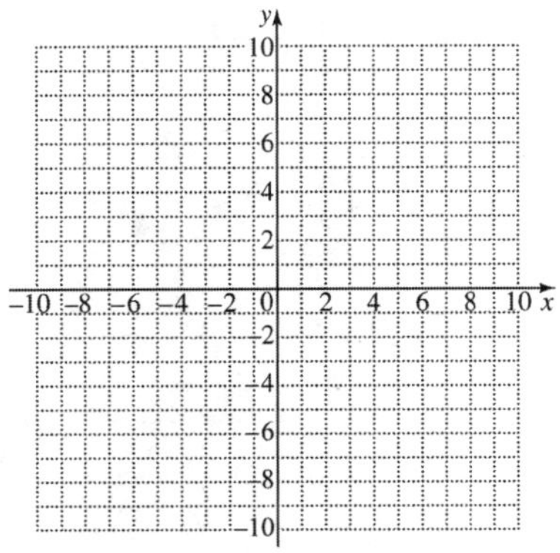

Solve the system using substitution.

2. $\begin{cases} 4x - 3y = 29 \\ 2x + y = 7 \end{cases}$

2. ______________________

Solve the system using elimination.

3. $\begin{cases} -3x + 5y = -13 \\ 4x - 2y = -6 \end{cases}$

3. ______________________

Name: **Date:**

Instructor: **Section:**

Solve each system using matrices.

4. $\begin{cases} x+3y=-2 \\ x-2y=3 \end{cases}$

4. ____________________

5. $\begin{cases} 2x+4y+z=4 \\ -3x-2y+3z=-17 \\ 4x-y-4z=8 \end{cases}$

5. ____________________

Solve the following systems using the method of your choice.

6. $\begin{cases} 10x-12y=6 \\ x=\frac{6}{5}y+\frac{3}{5} \end{cases}$

6. ____________________

Name: **Date:**

Instructor: **Section:**

7. $\begin{cases} 5x+3y=-1 \\ -3x-4y=-6 \end{cases}$

7. ______________________

8. $\begin{cases} x+3y-z=4 \\ 2x-4y+z=-10 \\ -3x+5y-4z=2 \end{cases}$

8. ______________________

9. $\begin{cases} x-2y=6 \\ 4x-8y=9 \end{cases}$

9. ______________________

Name: **Date:**

Instructor: **Section:**

Solve.

10. One number is 6 more than half the second number. The sum of the three times the first number and twice the second number is negative three.

10. ____________________

11. The sum of a three digit number is 15. The third digit is one more than twice the first digit. If you double the third digit the result is the same as the sum of the first two digits. What is the three digit number?

11. ____________________

12. A college was selling tickets to its annual fine arts department art show. The price for adults was \$8 and the price for children was \$5. If the department made \$572 dollars, and a total of 82 tickets were sold, how many on each type of ticket were sold?

12. ____________________

Name: **Date:**

Instructor: **Section:**

13. Given that $C(x) = 0.90x + 24{,}000$ and $R(x) = 2.10x$, find the break –even point.

13. ______________________

Graph the system of inequalities.

14. $\begin{cases} -x + 2y < 6 \\ 3x + 4y \geq 8 \end{cases}$

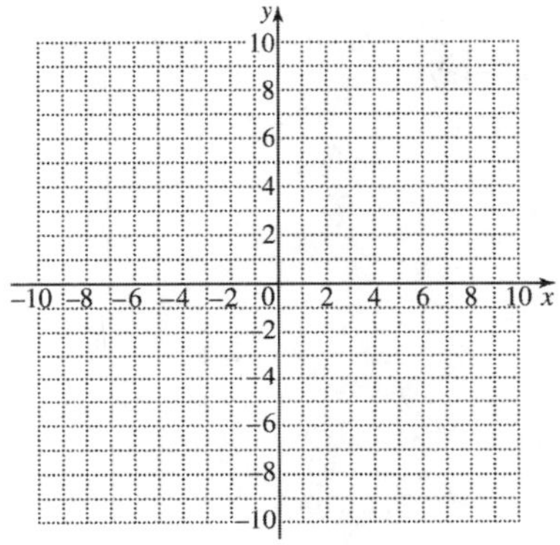

15. $\begin{cases} 3x + y < 5 \\ x - y > -4 \\ x \geq -3 \end{cases}$

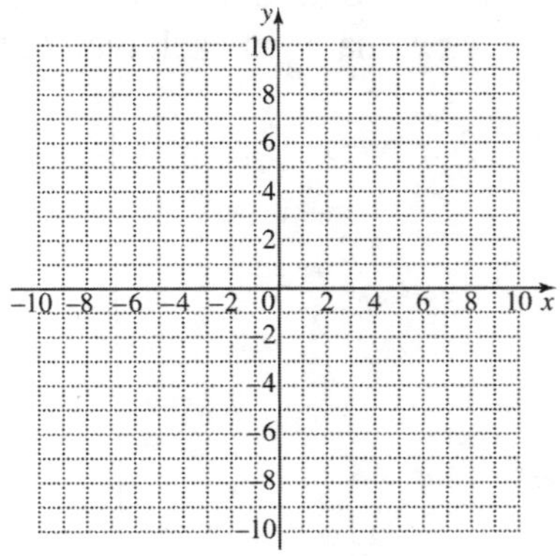

Name: **Date:**

Instructor: **Section:**

Practice Test B

Solve the system by substitution.

1. $\begin{cases} 2x+6y=8 \\ 3x-y=12 \end{cases}$ 1. ______________

a. $(1,2)$ b. $(5,3)$

c. $(4,0)$ d. $(2,-2)$

2. $\begin{cases} 5x=-3y+3 \\ 3y=4x-6 \end{cases}$ 2. ______________

a. $(0,1)$ b. $(1,-\frac{2}{3})$

c. $\varnothing$ d. $\{(x,y)\,|\,5x=-3y+3\}$

Solve the system using elimination.

3. $\begin{cases} \frac{1}{3}x-\frac{3}{4}y=-\frac{4}{3} \\ \frac{1}{2}x+\frac{3}{8}y=4 \end{cases}$ 3. ______________

a. $(\frac{1}{2},2)$ b. $(5,4)$

c. $\varnothing$ d. $\left\{(x,y)\,|\,\frac{1}{3}x-\frac{3}{4}y=-\frac{4}{3}\right\}$ $(2,-2)$

Name: **Date:**

Instructor: **Section:**

4. $\begin{cases} 2x - y = -1 \\ 8x + 10y = -18 \end{cases}$

4. ____________________

a. $(-1,-1)$ b. $(-2,-3)$

c. $\varnothing$ d. $\{(x,y) \mid 2x - y = -1\}$

Solve the system of equations using the method of your choice.

5. $\begin{cases} 2x - 3y + z = -18 \\ 3x + 4y + 2z = -4 \\ -2x - 5y - 3z = 4 \end{cases}$

5. ____________________

a. $(0,1)$ b. $(-2,3,-5)$

c. $\varnothing$ d. $\{(x,y,z) \mid 2x - 3y + z = 18\}$

6. $\begin{cases} 3x - 6y = 18 \\ -x + 2y = -6 \end{cases}$

6. ____________________

a. $(0,3)$ b. $(6,0)$

c. $\varnothing$ d. $\{(x,y) \mid 3x - 6y = 18\}$

Name: **Date:**

Instructor: **Section:**

7. $\begin{cases} x+2y+3z=7 \\ -2x-3y+z=24 \\ 3x+4y+2z=-13 \end{cases}$

7. ______________________

a. $(0,2,1)$ b. $(-3,-4,6)$

c. $\varnothing$ d. $\{(x,y,z)\mid x+2y+3z=7\}$

8. $\begin{cases} 3x+4y=8 \\ y=-\frac{3}{4}x-2 \end{cases}$

8. ______________________

a. $(0,2)$ b. $(-4,1)$

c. $\varnothing$ d. $\{(x,y)\mid 3x+4y=8\}$

Solve the system by graphing.

9. $\begin{cases} x+2y=6 \\ -3x+4y=-8 \end{cases}$

9. ______________________

a. $(4,1)$ b. $(0,3)$

c. $\varnothing$ d. $\{(x,y)\mid x+2y=6\}$

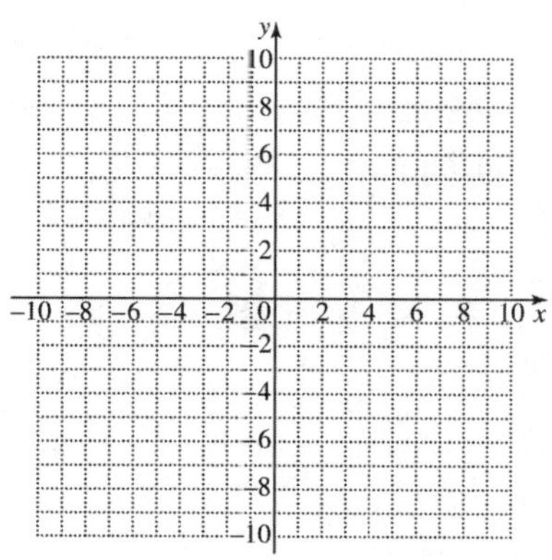

Name: **Date:**

Instructor: **Section:**

Solve.

10. Joe has 13 coins in his piggy bank. He only has three types of coins quarters, dimes, and nickels. If Joe doubles the number of nickels, he will have 2 more nickels than the total of dimes and quarters. Find how many of each coin Joe has if he has $1.35 in his piggy bank.

10. ____________________

a. 3 quarters, 4 dimes, 6 nickels b. 4 quarters, 5 dimes, 4 nickels

c. 3 quarters, 7 dimes, 3 nickels d. 2 quarters, 6 dimes, 5 nickels

11. The perimeter of a triangle is 24 inches. The first side is 3 more than the second side. The sum of the second and third side is two more than the first side. Find the three sides.

11. ____________________

a. 10, 7, and 7 b. 9, 6, and 9

c. 11, 8, and 5 d. 12, 9, and 3

12. The sum of a number and three times a number is four. The sum of three times the first number and nine times the second number is twelve. Find the numbers.

12. ____________________

a. 1 and 1 b. -2 and 2

c. $\varnothing$ d. $\{(x, y) \mid x + 3y = 4\}$

Name: **Date:**

Instructor: **Section:**

13. Find the measure of the angles of a triangle if the measure of one angle is twice the measure of the second angle and the third angle doubled is the same as the sum of the other two angles. Remember that the sum on the angles in a triangle total 180 degrees.

13. ____________________

a. $40°, 65°, 75°$ b. $30°, 70°, 80°$

c. $80°, 40°, 60°$ d. $50°, 60°, 70°$

14. Two cars left Chicago, IL traveling in opposite directions. After 4 hours they were 464 miles apart. If one car travels at 4 mph faster than the other car, find the rate of each car.

14. ____________________

a. 52 mph and 56 mph b. 58 mph and 62 mph

c. 56 mph and 60 mph d. 60 mph and 64 mph

15. Given the cost function $C(x) = 15x + 25000$ and the revenue function, $R(x) = 25x$. Find the break even point.

15. ____________________

a. 2000 units b. 2500 units

c. 625 units d. 250 units

Name: Date:
Instructor: Section:

Chapter 5 Exponents, Polynomials, and Polynomial Functions
Section 5.1 Exponents and Scientific Notation

Learning Objectives

1 .Use the product rule for exponents.
2. Evaluate expressions raised to the 0 power.
3. Use quotient rule for exponents.
4. Evaluate expressions raised to the *n*th power.
5. Convert between scientific notation and standard notation.

Objective 1

Use the product rule to simplify each expression.

1. $\left(3x^3y^4\right)\left(-5x^2y^3\right)$ 1. ____________________

2. $\left(-4x^3p^2\right)\left(4y^3x^3\right)$ 2. ____________________

3. $t^2 \cdot t \cdot t^6 \cdot \left(3t^4\right)$ 3. ____________________

Objective 2

Evaluate each expression.

4. -6^0 4. ____________________

5. $-5x^0$ 5. ____________________

Name: **Date:**
Instructor: **Section:**

6. $(5x^2z)^0$ 6. ____________

Objective 3

Use the quotient rule to simplify.

7. $\frac{t^5}{2t}$ 7. ____________

8. $\frac{5a^2b^4}{ab^2}$ 8. ____________

9. $\frac{48x^7y^9}{32x^4y^7}$ 9. ____________

Objective 4

Simplify and write using positive exponents only.

10. 6^{-2} 10. ____________

11. $\frac{a^2bc^3}{a^4b^5}$ 11. ____________

Name: **Date:**
Instructor: **Section:**

12. $\dfrac{1}{7^{-3}}$ 12. ____________

13. $\dfrac{14x^{-2}yz^{-4}}{2xyz}$ 13. ____________

14. $\dfrac{x^{3t-1}}{x^{t}}$ 14. ____________

15. $\dfrac{15az^{3}\left(4z^{-5}\right)}{30a^{-3}z^{4}}$ 15. ____________

Objective 5

Write each number in scientific notation.

16. 64,000,000 16. ____________

17. 0.000000456 17. ____________

Name:
Instructor:

Date:
Section:

Write each number in standard notation.

18. 3.26×10^{9}

18. ____________________

19. 2.3×10^{-13}

19. ____________________

Concept Extension

20. Prove the power of zero rule. (Hint: Use quotient rule)

20. ____________________

Name: **Date:**
Instructor: **Section:**

Section 5.2 More Work with Exponents and Scientific Notation

Learning Objectives

1. Use the power rule for exponents.
2. Use exponent rules and definitions to simplify exponential expressions.
3. Compute, using scientific notation.

Objective 1

Simplify. Write each answer using positive exponents only.

1. $\left(a^2b^{-3}c^{-4}\right)^{-3}$ 1. ____________

2. $\left(6^2\right)^{-2}$ 2. ____________

3. $\left(-2x^2y^0z^{-4}\right)^6$ 3. ____________

4. $\left(\dfrac{2t^{-3}}{s^4}\right)^3$ 4. ____________

Name: **Date:**
Instructor: **Section:**

Objective 2

Simplify. Write each answer using positive exponents.

5. $\left(\frac{2a^2y^{-2}}{y^{-3}}\right)^{-3}$

5. ____________________

6. $\frac{\left(3s^2t^4\right)^3}{15s^4t^8}$

6. ____________________

7. $\frac{\left(y^3\right)^{-4}}{y^3}$

7. ____________________

8. $\left(n^{x+4}\right)^2$

8. ____________________

Name: **Date:**
Instructor: **Section:**

9. $\dfrac{\left(2x^{2n+1}\right)^5}{x^{n+2}}$

9. ____________________

Objective 3

Perform each indicated operation. Write each answer in scientific notation.

10. $\left(5\times10^{11}\right)\left(2.9\times10^{-3}\right)$

10. ____________________

11. $\dfrac{3.6\times10^{5}}{0.4\times10^{-6}}$

11. ____________________

12. $\dfrac{66{,}000\times0.001}{0.002\times0.003}$

12. ____________________

Name: **Date:**
Instructor: **Section:**

Concept Extension

13. Is there such a number that if raised to the exponent of 1 is equivalent to that number being raised to the power of – 1?

13. ______________________

14. Is there such a number that if raised to the power of – 4, the result will be negative?

14. ______________________

Name: **Date:**
Instructor: **Section:**

Section 5.3 Polynomials and Polynomial Functions

Learning Objectives

1. Identify term, constant, polynomial, monomial, binomial, trinomial, and the degree of a term and of a polynomial.
2. Define polynomial function.
3. Review combining like terms.
4. Add polynomials.
5. Subtract polynomials.
6. Recognize the graph of a polynomial function from the degree of the polynomial.

Vocabulary

Use the choices to complete each statement.

Binomial	**Coefficient**	**Degree**
Like	**Monomial**	**Polynomial**
Term	**Trinomial**	**Unlike**
Variables		

1. A polynomial with three terms is called a ____________________________.
2. A ___________________ is a finite sum of terms in which each term is in the form ax^n, where n is a non-negative integer and a is a real number.
3. A polynomial with two terms is called a ________________________.
4. The degree of the polynomial is the largest _________________ of the terms in that polynomial.
5. If two terms have identical variables raised to identical powers, they are said to be _____________ terms.
6. A single term polynomial is called a ___________________.
7. The numerical factor of each term is its __________________________.
8. The degree of term is found by finding the sum of the exponents on the _____________________.

Objective 1

Find the degree of each term.

9. $35x^{12}$ 9. ______________________

10. -6 10. ______________________

Name: **Date:**
Instructor: **Section:**

11. $15x^{10}y^{12}z^{13}$ 11. ____________________

Find the degree of each polynomial and indicate whether the polynomial is a monomial, binomial, trinomial, or none of these.

12. $4x^2 + 2x - 1$ 12. ____________________

13. $5x^2y^2 - 4xy^2 + 6x^2y + 7y^2$ 13. ____________________

14. $15 - x^2$ 14. ____________________

Objective 2

If $P(x) = x^2 + x + 1$ and $Q(x) = 5x^2 + 1$, find the following.

15. $Q(-10)$ 15. ____________________

16. $P\left(\frac{1}{3}\right)$ 16. ____________________

Name: **Date:**
Instructor: **Section:**

Objective 3

Simplify by combining like terms.

17. $7x^2 - 2xy + 5y^2 - x^2 + xy + 11y^2$

17. ____________________

18. $t^2 + 2st - 2t^2 + 5st + s^2$

18. ____________________

Objective 4

Perform the indicated operation.

19. $(3y^2 + 6y - 12) + (5y^2 - 8y + 7)$

19. ____________________

20. $(4x^2 + 3xy - y^2) + (x^2 - 7xy + 5y^2)$

20. ____________________

21.

$$\begin{array}{r} 6t^2 + 5t - 10 \\ \underline{+(3t^2 - 10t + 15)} \end{array}$$

21. ____________________

Objective 5

Perform the indicated operation.

22. $(9x^3 - 2x^2 + 4x - 7) - (2x^3 - 6x^2 - 4x + 3)$

22. ______________________

23. Subtract $(x^2 - 6)$ from the sum of $(3x^2 - 5x + 6)$ and $(x^2 - 6x - 10)$.

23. ______________________

24.

$$\begin{array}{r} -5x^2 + 3x - 10 \\ \underline{-(3x^2 - 15x + 10)} \end{array}$$

24. ______________________

Concept Extension

For the given P(x) and Q(x) from # 15 and 16, find the following.

25. $2P(x) - 3Q(x)$

25. ______________________

26. $-3P(x) + 5Q(x)$

26. ______________________

Name: **Date:**
Instructor: **Section:**

Section 5.4 Multiplying Polynomials

Learning Objectives

1. Multiply two polynomials.
2. Multiply two binomials.
3. Square binomials.
4. Multiply the sum and difference of two terms.
5. Multiply three or more polynomials.
6. Evaluate polynomial functions.

Objective 1

Multiply.

1. $(2x^3)(-4x^2+7)$ 1. ____________

2. $(x+1)(x^2-3x+2)$ 2. ____________

3. $(x^2-2x+1)(x^2+3x+4)$ 3. ____________

Objective 2

Multiply the binomials.

4. $(3x-1)(x+3)$ 4. ____________

Name:
Instructor:

Date:
Section:

5. $(3a+b)(2a-5b)$

5. ____________________

6. $(2x-9)(x+3)$

6. ____________________

7. $(-x-1)(-5x+6)$

7. ____________________

Objective 3

Multiply, using special product methods.

8. $(3x+5)^2$

8. ____________________

9. $(4x^2-1)^2$

9. ____________________

10. $(5x^2-7)^2$

10. ____________________

Name: **Date:**
Instructor: **Section:**

Objective 4

Multiply, using special product rules.

11. $(5x-3)(5x+3)$ 11. ______

12. $(11x+8)(11x-8)$ 12. ______

13. $[(2s-3)-1][(2s-3)+1]$ 13. ______

Objective 5

Multiply.

14. $(x+y)(x-y)(x^2-y^2)$ 14. ______

15. $(y+3)(y-2)(y+1)$ 15. ______

Name: **Date:**
Instructor: **Section:**

16. $(r+1)^3$ 16. ____________

17. $(y-2)^4$ 17. ____________

Objective 6

If $f(x) = 3x^2 - 4x + 2$, find the following.

18. $f(-2)$ 18. ____________

19. $f(0)$ 19. ____________

20. $f(a)$ 20. ____________

Name: **Date:**
Instructor: **Section:**

Concept Extension

21. Using the function from # 18 – 20, find $\dfrac{f(a+h)-f(a)}{h}$.

21. ______________________

Name:
Instructor:

Date:
Section:

Section 5.5 The Greatest Common Factor and Factoring by Grouping

Learning Objectives
1. Identify the GCF.
2. Factor out the GCF of a polynomial's terms.
3. Factor polynomials by grouping.

Objective 1

Find the GCF of each list of monomials.

1. $10x^2yz^3,\ 20x^2z^5,\ 45xz^3$ — 1. ________

2. $24a^6b^4c,\ 36a^4b^6c^3,\ 48a^3b^7c$ — 2. ________

3. $16x^4y^7z^2,\ -24x^3y^5z^4,\ 18xy^6z^3$ — 3. ________

Objective 2

Factor out the GCF in each polynomial.

4. $15x-25y$ — 4. ________

5. $6x^5-8x^4+2x^3$ — 5. ________

Name: **Date:**
Instructor: **Section:**

6. $18a^4b^3 - 36a^3b^4 + 12a^2b^2$

6. ______________________

7. $2x(x-7) - 3(x-7)$

7. ______________________

8. $4x(x-1) - 8(x-1)$

8. ______________________

Objective 3

Factor by grouping.

9. $12xy - 8x - 3y + 2$

9. ______________________

10. $6x - 4y + 15x^2 - 10xy$

10. ______________________

11. $3 - 9b + 2a - 6ab$

11. ______________________

Name: **Date:**
Instructor: **Section:**

Concept Extension

Factor out the GCF of the following.

12. $4x^{6n} - 8x^{4n} + 16x^{2n}$

12. ____________________

Name: **Date:**
Instructor: **Section:**

Section 5.6 Factoring Trinomials

Learning Objectives

1. Factor trinomials of the form $x^2 + bx + c$.
2. Factor trinomials of the form $ax^2 + bx + c$.
    a. Method 1 – Trial and Check
    b. Method 2 - Grouping
3. Factor by substitution.

Objective 1

Factor each trinomial.

1. $x^2 + 2x - 8$ 1. ____________________

2. $x^2 - 8x + 15$ 2. ____________________

3. $x^2 + 4x - 12$ 3. ____________________

4. $x^2 + x - 20$ 4. ____________________

Name: **Date:**
Instructor: **Section:**

5. $3x^2 - 18x + 24$ 5. ____________________

Objective 2

Factor each trinomial.

6. $2x^2 - 3x - 2$ 6. ____________________

7. $25x^2 + 10x + 1$ 7. ____________________

8. $9x^2 + 18x + 8$ 8. ____________________

9. $4x^2 + 8x + 3$ 9. ____________________

10. $8a^2 - 2a - 1$ 10. ____________________

Name: **Date:**
Instructor: **Section:**

Objective 3

Use substitution to factor each polynomial completely.

11. $(x-4)^2+3(x-4)-18$

11. ____________________

12. $3x^4-3x^2-6$

12. ____________________

13. x^4-7x^2+10

13. ____________________

14. $10x^4+13x^2-3$

14. ____________________

15. $16x^4+24x^2+9$

15. ____________________

Name: Date:
Instructor: Section:

Concept Extension

16. Find all positive values of b such that $x^2 + bx + 8$ is factorable.

16. ____________________

17. Find all the negative values of b such that $x^2 + bx - 12$ is factorable.

17. ____________________

Name: Date:
Instructor: Section:

Section 5.7 Factoring by Special Products

Learning Objectives
1. Factor a perfect square trinomial.
2. Factor the difference of two squares.
3. Factor the sum or difference of two cubes.

Objective 1

1. $x^2+8x+16$ 1. ____________

2. $4x^2+20x+25$ 2. ____________

3. $4x^2-12x+9$ 3. ____________

4. $-72x^3-120x^2-50x$ 4. ____________

Objective 2

Factor the following.

5. x^2-81 5. ____________

Name: **Date:**
Instructor: **Section:**

6. $25t^2 - 36n^2$

6. ____________________

7. $-32x^2 + 50y^2$

7. ____________________

8. $16x^2 - 49y^2$

8. ____________________

Objective 3

Factor the following.

9. $z^3 - 1$

9. ____________________

10. $8x^3 + 343$

10. ____________________

11. $3x^3 + 24$

11. ____________________

Name: **Date:**
Instructor: **Section:**

12. $3x^6y^2 + 81y^2$ 12. ____________________

13. $a^6 - b^9$ 13. ____________________

Concept Extension

Factor completely.

14. $x^8 - 1$ 14. ____________________

15. $4x^{2n} - 49$ 15. ____________________

Name:
Instructor:

Date:
Section:

Section 5.8 Solving Equations by Factoring and Problem Solving

Learning Objectives

1. Solve polynomial equations by factoring.
2. Solve problems that can be modeled by polynomial equations.
3. Find the x-intercepts of a polynomial function.

Objective 1

Solve each equation.

1. $(x+5)(x-7)=0$ 1. ________________

2. $2x(x+1)(x-8)=0$ 2. ________________

3. $3x^2-3x-6=0$ 3. ________________

4. $2x^3+8x=42x$ 4. ________________

5. $4x^2-20x+16=0$ 5. ________________

Name: **Date:**
Instructor: **Section:**

6. $x^3 - x = 2x^2 - 2$

6. ____________________

7. $x(5x+2) = 3$

7. ____________________

8. $16x^2 + 56x = -49$

8. ____________________

9. $25y^2 - 36 = 0$

9. ____________________

Objective 2

Solve.

10. Determine whether any three consecutive integers represent the lengths of the sides of a right triangle.

10. ____________________

Name: **Date:**
Instructor: **Section:**

11. The sum of two numbers is 8. The sum of their squares is 34. Find the two numbers.

11. ____________________

12. An object falling off a cliff follows the path given by the equation $h(t) = -16t^2 + 4900$, where h(t) is the height at a given time, t, in seconds. At what time will the object hit the ground.

12. ____________________

13. If one number is 6 greater than another number, and their product is 91, find the two numbers.

13. ____________________

Objective 4

Find the x-intercepts of the following functions.

14. $y = x^2 - 5x + 6$

14. ____________________

Name: **Date:**
Instructor: **Section:**

15. $y = 6x^2 + x - 2$ 15. ____________________

16. $y = 9x^2 - 16$ 16. ____________________

17. $y = 4x^2 + 20x + 25$ 17. ____________________

Concept Extension

Solve.

18. $(x^2 - 6x + 8)(10x^2 + 13x + 3) = 0$ 18. ____________________

Name: **Date:**
Instructor: **Section:**

Chapter 5 Vocabulary

Vocabulary Word	Definition	Example
Product Rule	$x^m \cdot x^n = x^{m+n}$	$\left(a^2b\right)\left(a^3b^2\right) = a^5b^3$
Quotient Rule	$\frac{x^m}{x^n} = x^{m-n}$	$\frac{x^5}{x^3} = x^2$
Power Rule	$\left(x^m\right)^n = x^{m \cdot n}$	$\left(3^4\right)^5 = 3^{20}$
Zero Exponent	Anything to the zero power = 1	$5^0 = 1$
Polynomial	A finite sum of terms where each term is in the form ax^n ; n must be a non negative number.	$4x^2 - 3x + 7$
Monomial	One term polynomial	$5x$
Binomial	Two term polynomial	$2x - 7$
Trinomial	A three term polynomial	$4x^2 - 3x + 5$

Name: **Date:**
Instructor: **Section:**

Practice Test A

Simplify.

1. $\left(3x^4\right)\left(-6x^2\right)$ 1. ____________________

2. $\dfrac{r^{-5}s^4t}{rs^{-2}t^{-5}}$ 2. ____________________

3. $\left(2s^2t^{-4}u^3\right)^3$ 3. ____________________

4. $\dfrac{\left(a^{-2}b^3\right)^4 c^3}{b^{12}c^8}$ 4. ____________________

Write each number in scientific notation.

5. 0.00000038 5. ____________________

6. 4,500,000,000 6. ____________________

Name: **Date:**
Instructor: **Section:**

Use scientific notation to simplify each expression.

7. $(2.2\times10^{11})(3.0\times10^{-16})$ 7. ______________

8. $\dfrac{(0.0000042)(1{,}500{,}000)}{6{,}000{,}000}$ 8. ______________

Perform the indicated operation.

9. $(3x^2-10x+5)+(5x^2+12x-7)$ 9. ______________

10. $(4x^3+5x^2-3x+7)-(3x^2-6x-13)$ 10. ______________

11. $(3x-5)(4x+7)$ 11. ______________

Name: **Date:**
Instructor: **Section:**

12. $(15x+4)^2$ — 12. ____________________

13. $(18x-13)(18x+13)$ — 13. ____________________

14. $(2x+3)(x^2-3x+4)$ — 14. ____________________

Factor each polynomial completely.

15. $x^2-8x+15$ — 15. ____________________

16. $8x^3-64$ — 16. ____________________

Name: **Date:**
Instructor: **Section:**

17. $5c^2 - 15c + 3ac - 9a$ 17. ______________________

18. $36x^2 + 24x + 4$ 18. ______________________

19. $x^2 + 81$ 19. ______________________

Solve the equation for the variable.

20. $x^2 - 16 = 0$ 20. ______________________

21. $x^2 + 20 = 9x$ 21. ______________________

Name: **Date:**
Instructor: **Section:**

22. $(x-5)(2x+1)=0$

22. ______________________

23. $8x^3-32x=0$

23. ______________________

24. $4x^2-12x+9=0$

24. ______________________

Find the x – intercept of the given function.

25. $y=2x^2-x-3$

25. ______________________

Name:

Instructor:

Date:

Section:

Practice Test B

Simplify.

1. $2x^3(3x^2 - 7x)$ 1. ____________________

a. $5x^5 - 9x^4$ b. $5x^6 - 5x^4$

c. $6x^5 - 14x^4$ d. $6x^6 - 14x^3$

2. $\dfrac{a^{-5}b^6}{a^8b^{-4}}$ 2. ____________________

a. $\dfrac{b^{10}}{a^{13}}$ b. $\dfrac{a^{13}}{b^{10}}$

c. $a^{13}b^{10}$ d. a^3b^2

3. $(5c^{-4}d^2)^{-3}$ 3. ____________________

a. $\dfrac{125c^{12}}{d^6}$ b. $\dfrac{c^{12}}{125d^6}$

c. $\dfrac{d^6}{125c^{12}}$ d. $\dfrac{125}{c^7d}$

4. $\dfrac{(x^5y^6)^2}{x^{10}y^{14}}$ 4. ____________________

a. $\dfrac{1}{x^3y^2}$ b. $\dfrac{x}{y^2}$

c. $\dfrac{1}{y^2}$ d. y^2

Name: **Date:**
Instructor: **Section:**

Write in scientific notation.

5. 0.000000046 5. ____________________

a. 4.6×10^{-8} b. 46×10^{9}

c. 46×10^{-9} d. 4.6×10^{8}

Write in scientific notation.

6. $34,700,000,000$ 6. ____________________

a. 34.7×10^{-9} b. 34.7×10^{9}

c. 3.47×10^{-8} d. 3.47×10^{10}

Use scientific notation to find the following.

7. $\dfrac{1.4\times10^{-11}}{2.8\times10^{-18}}$ 7. ____________________

a. 0.5×10^{7} b. 5×10^{6}

c. 2×10^{8} d. 2×10^{-7}

8. $(1,890,000,000)(0.000000014)$ 8. ____________________

a. 2.646×10 b. 26.46×10^{-2}

c. 264.6 d. 2.646×10^{-1}

Name: **Date:**
Instructor: **Section:**

Perform the indicated operation.

9. $(3x+5)+(2x^2-5x+7)$ 9. ____________

a. $2x^2-2x+12$ b. $2x^2+8x+12$

c. $2x^2-8x-2$ d. $2x^2-2x-2$

10. Subtract $(6x-5)$ from the sum of $(10x+7)$ and (x^2-5x+6).

10. ____________

a. $x^2+11x+8$ b. x^2-x+18

c. x^2-x+8 d. $x^2+11x+18$

11. $(5x+7)(3x-8)$ 11. ____________

a. $8x^2+8x-56$ b. $15x^2-19x-56$

c. $8x^2+19x-56$ d. $15x^2+8x-56$

12. $(3x-2)(2x^2+5x-8)$ 12. ____________

a. $6x^3+11x^2-34x+16$ b. $6x^3+11x^2+14x-16$

c. $6x^3+19x^2+34x+16$ d. $6x^3-11x^2+14x+16$

Name: **Date:**
Instructor: **Section:**

13. $(4x-1)^3$

13. ____________________

a. $64x^3+16x^2-4x-1$

b. $64x^3+48x^2+12x+1$

c. $64x^3-16x^2-4x+1$

d. $64x^3-48x^2+12x-1$

14. $(12x-11y)(12x+11y)$

14. ____________________

a. $144x^2-121y^2$

b. $144x^2+264xy+121y^2$

c. $24x^2-22y^2$

d. $144x^2+121y^2$

Factor completely.

15. $3x^3-81$

15. ____________________

a. $3(x-3)(x-3x+9)$

b. $(3x-9)(x^2+9)$

c. $3(x+3)(x-3x+9)$

d. $3(x-3)(x^2+3x+9)$

16. x^2+5x+6

16. ____________________

a. $(x-2)(x-3)$

b. $(x-1)(x+6)$

c. $(x+2)(x+3)$

d. $(x+1)(x-6)$

Name:
Instructor:

Date:
Section:

17. $8x^2+2x-3$

17. ________________

a. $(4x-3)(2x+1)$

b. $(4x+3)(2x-1)$

c. $(8x-3)(x+1)$

d. prime

18. $25p^2-36$

18. ________________

a. $(5p+6)^2$

b. $(5p-6)^2$

c. $(5p+6)(5p-6)$

d. prime

Solve the equation for the variable.

19. $x^2-16x+15=0$

19. ________________

a. 15 and 1

b. – 15 and – 1

c. 3 and – 5

d. – 15 and 1

20. $4x^2-20x+16=0$

20. ________________

a. 16 and 1

b. 4 and 1

c. – 4 and – 1

d. – 16 and 1

21. $3x^3-21x^2=-30x$

21. ________________

a. 15, 2, and 0

b. 6, 5, and 1

c. – 5, 6, and 1

d. 5, 2, and 0

Name: **Date:**
Instructor: **Section:**

22. $4a^2 - 64$ 22. ____________

a. -4 and 4 b. 4

c. -4 d. 2, - 2

23. $3x^2 = x + 2$ 23. ____________

a. 1 and $\frac{2}{3}$ b. 1 and 2

c. 1 and $-\frac{2}{3}$ d. $\frac{1}{3}$ and 2

Find the x-intercept of the following equations.

24. $y = 8x^3 + 10x^2 - 3x$ 24. ____________

a. $\frac{3}{8}$, 1, 0 b. $\frac{1}{4}, -\frac{3}{2}$ and 0

c. $-\frac{3}{8}$, -1, 0 d. 3, 1, $\frac{1}{8}$

25. $y = 2x^2 - 12x$ 25. ____________

a. 12 and 1 b. 12 and -1

c. -6 and 2 d. 6 and 0

Name:
Instructor:

Date:
Section:

Chapter 6 Rational Expressions

Section 6.1 Rational Functions and Multiplying and Dividing Rational Expressions

Learning Objectives

1. Find the domain of a rational expression.
2. Simplify rational expressions.
3. Multiply rational expressions.
4. Divide rational expressions.
5. Use rational functions in applications.

Vocabulary

Use the choices to complete each statement.

Domain	**Rational**	**Simplified**
1	**-1**	**0**

1. The expression $\frac{x-y}{y-x}$ will simplify down to ____________________.
2. Like a fraction, if the denominator of a rational expression is ______________ then the expression in undefined.
3. The ______________ of the function $f(x)=\frac{6}{x+4}$ is $\{x\,|\,x \text{ is a real number}, x \neq -4\}$.
4. A __________________ expression is an expression that can be written as a quotient of two polynomials, but the denominator cannot be 0.
5. The expression $\frac{a+b}{b+a}$ will simplify down to ____________________.

Objective 1

Find the domain of each rational expression.

6. $f(x)=\frac{3x+1}{x-5}$

6. ____________________

7. $g(x)=\frac{4x}{x^2+3x+2}$

7. ____________________

Name: **Date:**
Instructor: **Section:**

8. $h(x) = \dfrac{5x}{3x+2}$

8. ______________________

9. $C(x) = \dfrac{x+3}{x^2-4}$

9. ______________________

Objective 2

Simplify each rational expression.

10. $\dfrac{6y-18}{8y-24}$

10. ______________________

11. $\dfrac{x^2+6x-40}{x+10}$

11. ______________________

12. $\dfrac{-a+b}{b-a}$

12. ______________________

13. $\dfrac{x^2-9}{x^2+x-6}$

13. ______________________

Name:
Instructor:

Date:
Section:

Objective 3

Multiply and simplify.

14. $\dfrac{x+2}{x-6} \bullet \dfrac{x-6}{x+4}$

14. ____________________

15. $\dfrac{x^2+5x-6}{x-3} \bullet \dfrac{x^2+x-12}{x^2-1}$

15. ____________________

16. $\dfrac{a^3-8}{a^2-4} \bullet \dfrac{a+2}{a^2+2a+4}$

16. ____________________

17. $\dfrac{5x^2+5}{x^2+4x+4} \bullet \dfrac{x^2+3x+2}{x^2-1}$

17. ____________________

Name: **Date:**
Instructor: **Section:**

Objective 4

Divide and simplify.

18. $\frac{4}{5x} \div \frac{4x^2 - 12}{10x^2 + 15x}$

18. ____________________

19. $\frac{3x - x^2}{x^3 - 27} \div \frac{x}{x^2 + 3x + 9}$

19. ____________________

20. $\frac{x^2 + 4x + 3}{2x^2 + 5x + 2} \div \frac{x^2 + 5x + 6}{2x^2 + 3x + 1}$

20. ____________________

Objective 5

21. The total revenue from the sale of a popular book is approximated by the rational expression $R(x) = \frac{1000x^2}{3x - 2}$, where x is the number of years since publication and $R(x)$ is the total revenue in millions of dollars.

a. Find the total revenue at the end of the first year.

21a. ____________________

b. Find the total revenue at the end of the second year.

21b. ____________________

c. Find the revenue during the second year only.

21c. ____________________

d. Find the domain of function R.

21d. ____________________

Name:
Instructor:

Date:
Section:

Concept Extension

22. Find the area of a triangle with height of $\frac{x}{x+3}$ and base of $\frac{x^2+6x+9}{2x}$.

22. ____________________

Name: Date:
Instructor: Section:

Section 6.2 Adding and Subtracting Rational Expressions

Learning Objectives

1. Add or subtract rational expressions with common denominators.
2. Identify the least common denominator of two or more rational expressions.
3. Add or subtract rational expressions with unlike denominators.

Objective 1

Add or subtract as indicated.

1. $\frac{x-7}{3x}-\frac{x+4}{3x}$ 1. ____________

2. $\frac{x-5}{x+1}+\frac{x-3}{x+1}$ 2. ____________

3. $\frac{x-5}{2x}-\frac{x+5}{2x}$ 3. ____________

4. $\frac{3x-4}{x^2+3x+4}-\frac{5x+3}{x^2+3x+4}$ 4. ____________

Name:
Instructor:

Date:
Section:

Objective 2

Find the LCD of the rational expressions in each list.

5. $\frac{7}{3x}, \frac{16}{9x^2}, \frac{5}{12x}$

5. ____________________

6. $\frac{x+1}{x-2}, \frac{x-1}{x^2-4}, \frac{x+3}{x+2}$

6. ____________________

7. $\frac{3x+1}{x}, \frac{x-2}{x+2}; \frac{x}{x-1}$

7. ____________________

8. $\frac{3x}{x+1}, \frac{2x+1}{x^2+3x+2}, \frac{x-4}{x^2+4x+4}$

8. ____________________

9. $\frac{2a}{a^2-b^2}, \frac{1}{a^2-2ab+b^2}$

9. ____________________

Name: **Date:**
Instructor: **Section:**

Objective 3

Add or subtract as indicated.

10. $\dfrac{12}{5x}+\dfrac{6}{25x^2}$

10. ____________________

11. $\dfrac{x-3}{x+2}-\dfrac{x-4}{x-2}$

11. ____________________

12. $\dfrac{y+1}{y^2-6y+8}-\dfrac{3}{y^2-16}$

12. ____________________

13. $\dfrac{y+1}{y^2+6y+8}-\dfrac{y-1}{y^2+7y+12}$

13. ____________________

14. $\dfrac{x+2}{2x^2+7x+3}+\dfrac{x+4}{2x^2-5x-3}$

14. ____________________

Name: **Date:**
Instructor: **Section:**

Concept Extension

15. Explain in your own words, how fractions are similar to rational expressions.

15. ____________________

16. Find the perimeter of a rectangle whose length is $\frac{x+2}{x^2-25}$ and width is $\frac{x-8}{x^2+10x+25}$.

16. ____________________

Name: **Date:**
Instructor: **Section:**

Section 6.3 Simplifying Complex Fractions

Learning Objective

1. Simplify complex fractions by simplifying the numerator and denominator and then dividing.
2. Simplify complex fractions by multiplying by a common denominator.
3. Simplify expressions with negative exponents.

Objective 1

Simplify each complex fraction.

1. $\dfrac{\dfrac{2}{5x}}{\dfrac{4}{15x}}$ 1. ______________

2. $\dfrac{\dfrac{3}{x+2}}{\dfrac{5}{x-2}}$ 2. ______________

3. $\dfrac{\dfrac{4x^2-y^2}{xy}}{\dfrac{2}{y}-\dfrac{1}{x}}$ 3. ______________

Name:
Instructor:

Date:
Section:

4. $\dfrac{\dfrac{4}{x}-\dfrac{3}{x^2}}{\dfrac{6}{x^3}+\dfrac{4}{x^2}}$

4. ____________________

Objective 2

Simplify each complex fraction.

5. $\dfrac{3-\dfrac{4}{y}}{2+\dfrac{2}{y}}$

5. ____________________

6. $\dfrac{\dfrac{x+2}{x}-\dfrac{2}{x-1}}{\dfrac{x+1}{x}+\dfrac{x+1}{x-1}}$

6. ____________________

7. $\dfrac{\dfrac{2}{xy}-4}{\dfrac{5}{x}+\dfrac{3}{y}}$

7. ____________________

Name: **Date:**
Instructor: **Section:**

8. $\dfrac{\dfrac{-5x}{x+y}}{\dfrac{x}{y^2}}$

8. ______________________

Objective 3

Simplify.

9. $\dfrac{2a^{-1}+3b^{-2}}{a^{-1}-b^{-1}}$

9. ______________________

10. $\dfrac{4+y^{-1}}{3y^{-1}-2y^{-2}}$

10. ______________________

11. $\dfrac{5-(3x)^{-2}}{3+x^{-3}}$

11. ______________________

Name: **Date:**
Instructor: **Section:**

Concept Extension

Simplify.

12. $\dfrac{(x-1)^{-2}+3x^{-1}}{5x^{-2}+(x-1)^{-2}}$

12. ____________________

Name: **Date:**
Instructor: **Section:**

Section 6.4 Dividing Polynomials: Long Division and Synthetic Division

Learning Objectives

1. Divide a polynomial by a monomial.
2. Divide by a polynomial
3. Use synthetic division to divide a polynomial by a binomial.
4. Use the remainder theorem to evaluate polynomials.

Objective 1

Divide.

1. $12x^4 + 24x^3 - 36x^2$ by $6x^2$. 1. ______________

2. $\dfrac{16x^2y^4 + 24xy^3 - 8x^2y^2}{8x^3y^3}$ 2. ______________

3. $15x^6 - 25x^4 + 30x^2$ by $15x^2$ 3. ______________

Objective 2

Divide.

4. $(x^2 + 16x + 15) \div (x + 15)$ 4. ______________

Name: **Date:**
Instructor: **Section:**

5. $\left(6x^2-13x-5\right)\div(2x-5)$ 5. ____________________

6. $\left(3x^5-x^3+4x^2-12x-8\right)\div\left(x^2-2\right)$ 6. ____________________

7. $\left(2x^3+x^2-35x+40\right)\div(x+5)$ 7. ____________________

Objective 3

Use synthetic division to divide.

8. $\left(x^2+7x+6\right)\div(x+6)$ 8. ____________________

9. $\dfrac{x^3-7x^2-13x+5}{x-2}$ 9. ____________________

Name: **Date:**
Instructor: **Section:**

10. $\dfrac{x^3+27}{x+3}$ 10. ______________

11. $\left(2x^4+\dfrac{1}{2}x^3+x^2+x\right)\div(x-2)$ 11. ______________

Objective 3

For the given polynomial $P(x)$ and the given c, use the remainder theorem to find $P(c)$.

12. $P(x)=4x^4+x^2-2;\ -1$ 12. ______________

13. $P(x)=4x^4+2x^3-6x+7;\ -2$ 13. ______________

14. $P(x)=5x^4-4x^3-7x^2-5x+1;\ 4$ 14. ______________

Name: **Date:**
Instructor: **Section:**

Concept Extension

Divide.

15. $\left(5x^4 + x^3 - x^2 + x - 1\right) \div \left(2x + 1\right)$

15. ____________________

Name: **Date:**
Instructor: **Section:**

Section 6.5 Solving Equations Containing Rational Expressions

Learning Objectives

1. Solve equations containing rational expressions.

Objective 1

Solve each equation.

1. $\frac{2}{x}+\frac{1}{2}=\frac{5}{x}$

1. ______________________

2. $\frac{x^2-5x}{7x}=\frac{6}{7x}$

2. ______________________

3. $\frac{6}{x+2}=\frac{4}{x-2}$

3. ______________________

Name:
Instructor:

Date:
Section:

4. $\dfrac{2x+1}{4-x}=\dfrac{9}{4-x}$

4. ____________________

5. $\dfrac{2}{x-3}+\dfrac{3}{x+4}=\dfrac{4}{x^2+x-12}$

5. ____________________

6. $\dfrac{x^2-20}{x^2-7x+12}=\dfrac{3}{x-3}+\dfrac{5}{x-4}$

6. ____________________

7. $\dfrac{5}{2x-5}-\dfrac{4}{3x-1}=\dfrac{2}{6x^2-17x+5}$

7. ____________________

Name: **Date:**
Instructor: **Section:**

8. $\frac{14}{2x^2 - 10x} + 1 = \frac{3}{x-5}$

8. ____________________

Concept Extension

Solve each equation.

9. $x^{-2} - 14x^{-1} + 24 = 0$

9. ____________________

10. $\left(\frac{2}{x+1}\right)^2 + 3\left(\frac{2}{x+1}\right) - 28 = 0$

10. ____________________

Name:
Instructor:

Date:
Section:

Section 6.6 Rational Equations and Problem Solving

Learning Objectives
1. Solve an equation containing rational expressions for a specified variable.
2. Solve problems by writing equations containing rational expressions.

Objective 1

Solve each equation for the specified variable.

1. $\frac{1}{p}+\frac{1}{f}=\frac{1}{c}$ for f.

1. ________________

2. $F=\frac{-GMm}{r^2}$ for G.

2. ________________

3. $A=\frac{h(a+b)}{2}$ for a.

3. ________________

4. $\frac{1}{R}=\frac{1}{R_1}+\frac{1}{R_2}$ for R.

4. ________________

Name: **Date:**
Instructor: **Section:**

Objective 2

Solve.

5. The sum of a number and 7 times its reciprocal is 8. Find the number.

5. ____________________

6. If a number is added to the denominator of $\frac{11}{3}$ and subtracted from the numerator, the resulting fraction is equivalent to $\frac{3}{4}$.

6. ____________________

7. Three postal workers can sort a stack of mail in 20 minutes, 30 minutes, and 60 minutes, respectively. Find how long it takes them to sort the mail if all three work together.

7. ____________________

Name: **Date:**
Instructor: **Section:**

8. Sue can grade tests in 2 hours, if she gets help from her husband, it will take them 45 minutes. how long would it take her husband to grade the tests for her?

8. ____________________

9. The speed of Lazy River's current is 5 mph. If a boat travels 20 miles downstream in the same time that it takes to travel 10 miles upstream, find the speed of the boat in still water.

9. ____________________

10. Two cars going in opposite directions leave at the same time. One car travels 10 mph faster then the other. In 5 hours they are 660 miles apart. Find the speed of each car.

10. ____________________

Name: **Date:**
Instructor: **Section:**

Section 6.7 Variation and Problem Solving

Learning Objective

1. Solve problems involving direct variation.
2. Solve problems involving inverse variation.
3. Solve problems involving joint variation.
4. Solve problems involving combined variation.

Objective 1

If y varies directly as x, find the constant of variation and the direct variation equation for each situation.

1. $y=6$ when $x=42$ 1. ____________________

2. $y=10$ when $x=80$ 2. ____________________

3. $y=16$ when $x=20$ 3. ____________________

4. $y=0.6$ when $x=1.8$ 4. ____________________

5. $y=\frac{3}{4}$ when $x=\frac{2}{3}$ 5. ____________________

Name:
Instructor:

Date:
Section:

Solve.

6. The weight of a synthetic ball varies directly with the cube of its radius. A ball with a radius of 2 inches weighs 1.20 pounds. Find the weight of a ball of the same material with a 3-inch radius.

6. ______________________

Objective 2

If y varies inversely as x, find the constant of variation and the inverse variation equation for each situation.

7. $y = 4$ when $x = 6$

7. ______________________

8. $y = 7$ when $x = 2$

8. ______________________

9. $y = \frac{1}{2}$ when $x = 8$

9. ______________________

Name: **Date:**
Instructor: **Section:**

10. $y = 0.3$ when $x = 0.2$

10. ______________________

Solve.

11. Pairs of markings a set distance apart are made on highways so that the police can detect drivers exceeding the speed limit. Over a fixed distance, the speed R varies inversely with the time T. In one particular pair of markings, R is 45 mph when T is 6 seconds. Find the speed of a car that travels the given distance in 5 seconds.

11. ______________________

Objective 3

Write each statement as an equation. Use k as the constant of variation.

12. y varies jointly as x and w.

12. ______________________

13. y varies jointly as the square of x and z; $y = 120$ when $x = 10$ and $z = 2$.

13. ______________________

14. y varies jointly as x and z cubed; $y = 320$ when $x = 20$ and $z = 2$.

14. ______________________

Name: **Date:**
Instructor: **Section:**

Solve.

15. The maximum weight that a rectangular beam can support varies jointly as its width and the square of its height and inversely as its length. If a beam $\frac{1}{2}$ foot wide, $\frac{1}{3}$ foot high, and 10 feet long can support 12 tons, find how much a similar beam can support if the beam is $\frac{2}{3}$ foot wide, $\frac{1}{2}$ foot high, and 16 feet long.

15. ____________________

Concept Extension

16. Suppose that y varies directly as x^3. If x is doubled, what is the effect on y?

16. ____________________

17. Suppose that y varies inversely as x^2. If x is doubled, what is the effect on y?

17. ____________________

Name: **Date:**
Instructor: **Section:**

Chapter 6 Vocabulary

Vocabulary Word	**Definition**	**Example**
Rational Expression	The quotient of two polynomials.	$\frac{x+3}{x-7}$
Rational Function	A function that can be described by using a rational expression.	$f(x)=\frac{2x+7}{x-9}$
Synthetic Division	A short cut method to divide a polynomial by a binomial of the form $x-c$	
Extraneous Solution	A solution that will cause the denominator to be equal to 0.	$\frac{x}{x-2}=\frac{2}{x-2}$ 2 is an extraneous solution.
y varies directly as x.	$y=kx$	$y=4x$
y varies inversely as x.	$y=\frac{k}{x}$	$y=\frac{24}{x}$
y varies jointly as x and z.	$y=kxz$	$y=3xz$

Name: **Date:**
Instructor: **Section:**

Practice Test A

Find the domain of the rational expression.

1. $f(x)=\frac{4x}{3-x}$ 1. ______________

2. $g(x)=\frac{3x+6}{x^2-9}$ 2. ______________

Write each rational expression in lowest terms.

3. $\frac{8x+32}{x^2+6x+8}$ 3. ______________

4. $\frac{x-3}{x^3-27}$ 4. ______________

5. $\frac{x^2-4x}{x^2-10x+24}$ 5. ______________

Name: **Date:**
Instructor: **Section:**

Perform the indicated operation. If possible, simplify your answer.

6. $\dfrac{3x}{x^2+7x+12} \bullet \dfrac{x^2+8x+16}{15x}$ 6. ____________

7. $\dfrac{x+5}{x^2-5x-50} \div \dfrac{x+2}{x^2-12x+20}$ 7. ____________

8. $\dfrac{2}{x+5} + \dfrac{6}{x-5}$ 8. ____________

9. $\dfrac{x+1}{x^2+6x+8} \bullet \dfrac{x+4}{x^2-1} \div \dfrac{x+5}{x-1}$ 9. ____________

Name:
Instructor:

Date:
Section:

10. $\dfrac{5x}{x-2}-\dfrac{3x}{x+2}$

10. ____________________

11. $\dfrac{8}{x^2-4}+\dfrac{2x}{x^2+4x+4}-\dfrac{6}{x+2}$

11. ____________________

12. $\dfrac{2x+5}{x^2+5x-6}\cdot\dfrac{x^2+4x-12}{x-1}$

12. ____________________

13. $\dfrac{x}{5}\left(\dfrac{4x}{3}-\dfrac{7}{6x}\right)$

13. ____________________

Name: **Date:**
Instructor: **Section:**

Simplify each complex fraction.

14. $\dfrac{\dfrac{1}{x}+\dfrac{4}{2x}}{\dfrac{3}{4x}-\dfrac{x}{2}}$

14. ____________________

15. $\dfrac{\dfrac{x^2+5x-6}{x+2}}{\dfrac{x^2+8x+12}{2x-1}}$

15. ____________________

Divide.

16. $\left(16x^3+12x^2-24x+20\right)\div 4x$

16. ____________________

17. $\dfrac{2x^2+7x+6}{x+2}$

17. ____________________

18. $(x^3+2x^2-x+1)\div(x-1)$

18. ____________________

19. If $P(x)=4x^3-6x^2+x-12$, use the remainder theorem to find $P(-1)$.

19. ____________________

Find the constant of variation and the variation equation.

20. y varies directly with the square of x; $y=2400$ when $x=20$.

20. ____________________

Name: **Date:**
Instructor: **Section:**

Practice Test B

Find the domain of the rational expression.

1. $f(x)=\dfrac{-2x+5}{5+x}$ 1. ____________________

a. $\{x \mid x \text{ is a real number}, x \neq \tfrac{5}{2}\}$ b. $\{x \mid x \text{ is a real number}, x \neq -\tfrac{5}{2}\}$

c. $\{x \mid x \text{ is a real number}, x \neq -5\}$ d. $\{x \mid x \text{ is a real number}, x \neq 5\}$

2. $g(x)=\dfrac{2x^2+4}{x^2+5x-6}$ 2. ____________________

a. $\{x \mid x \text{ is a real number}, x \neq -6, x \neq 1\}$ b. $\{x \mid x \text{ is a real number}, x \neq 6, x \neq -1\}$

c. $\{x \mid x \text{ is a real number}, x \neq -2, x \neq 2\}$ d. $\{x \mid x \text{ is a real number}, x \neq 2\}$

Write each rational expression in lowest terms.

3. $\dfrac{2x+4}{x^2+4x+4}$ 3. ____________________

a. $\dfrac{x+2}{x-2}$ b. 1

c. $\dfrac{2}{x-2}$ d. $\dfrac{2}{x+2}$

Name: **Date:**
Instructor: **Section:**

4. $\dfrac{x+1}{x^3+1}$ 4. ____________________

a. $\dfrac{1}{x^2+x+1}$ b. $\dfrac{x}{x^2-x+1}$

c. $\dfrac{1}{x^2-x+1}$ d. $\dfrac{x+1}{x^2-x+1}$

5. $\dfrac{x^2-5x}{x^2-10x+25}$ 5. ____________________

a. $\dfrac{x-5}{x+5}$ b. $\dfrac{x+5}{x}$

c. $\dfrac{x}{x-5}$ d. $\dfrac{x}{x+5}$

Perform the indicated operation. If possible, simplify your answer.

6. $\dfrac{x+2}{4x^2-16} \cdot \dfrac{x^2-7x-10}{x-5}$ 6. ____________________

a. 1 b. $\dfrac{1}{4}$

c.- 1 d. 4

Name: **Date:**
Instructor: **Section:**

7. $\dfrac{x-3}{x^2-2x-24} \div \dfrac{x^2-7x+12}{x+4}$

7. ____________________

a. $\dfrac{1}{(x-6)(x-4)}$

b. $\dfrac{x-6}{x-4}$

c. $\dfrac{x-3}{x+4}$

d. $\dfrac{(x-4)(x-3)}{(x+4)(x-6)}$

8. $\dfrac{4}{2x+7} - \dfrac{3}{2x-7}$

8. ____________________

a. $\dfrac{2x-7}{4x^2-49}$

b. $\dfrac{1}{2x}$

c. $\dfrac{1}{2x+7}$

d. $\dfrac{2x-49}{4x^2-49}$

9. $\dfrac{2x-1}{x^2+6x+9} \cdot \dfrac{x^2+2x-3}{x^2-2x+1} \div \dfrac{4x^2-1}{x-1}$

9. ____________________

a. $\dfrac{1}{(x+3)(2x-1)}$

b. $\dfrac{1}{(x+3)(2x+1)}$

c. $\dfrac{x+3}{2x-1}$

d. $\dfrac{(2x+1)^2(2x-1)}{(x+3)(x-1)^2}$

Name: **Date:**
Instructor: **Section:**

10. $\dfrac{4x}{x-7}-\dfrac{7}{x+7}$ 10. ______________

a. $\dfrac{4x-7}{2x}$ b. $\dfrac{4x^2+21x+49}{(x-7)(x+7)}$

c. $\dfrac{4x-7}{-14}$ d. $\dfrac{-3x+77}{(x-7)(x+7)}$

11. $\dfrac{5}{x+3}+\dfrac{3x}{x^2+6x+9}-\dfrac{8}{x^2-9}$ 11. ______________

a. $\dfrac{8x^2-17x-21}{(x+3)(x+3)(x-3)}$ b. $\dfrac{8x^2-17x-69}{(x+3)(x+3)(x-3)}$

c. $\dfrac{3x^2-17x-19}{(x+3)(x-3)}$ d. $\dfrac{3x^2-17x+29}{(x+3)(x-3)}$

12. $\dfrac{3x+4}{3x^2-2x-8}\cdot\dfrac{x^2-4}{x+2}$ 12. ______________

a. $\dfrac{x-2}{x+2}$ b. 1

c. $\dfrac{x+2}{x-2}$ d. -1

Name: **Date:**
Instructor: **Section:**

13. $\frac{-3}{x}\left(\frac{3}{4}-\frac{5}{8x}\right)$ 13. ____________________

a. $\frac{-18x+15}{8x^2}$ b. $\frac{6x-5}{24}$

c. $\frac{8x^2}{15-18x}$ d. $\frac{24}{6x-5}$

Simplify each complex fraction.

14. $\dfrac{\frac{x}{5}+\frac{3}{4x}}{\frac{x}{10}-\frac{5}{2x}}$ 14. ____________________

a. $\frac{4x^2+15}{2(x-5)(x+5)}$ b. $\frac{4x^2-11}{2(x+25)}$

c. $\frac{4x^2-15}{2(x+25)}$ d. $\frac{2(x-15)}{4x^2+15}$

15. $\dfrac{\frac{x^3+1}{x+4}}{\frac{x^2-x+1}{x^2+9x+20}}$ 15. ____________________

a. $\frac{x^2-x+1}{x-1}$ b. $\frac{x+5}{x+1}$

c. $(x+1)(x+5)$ d. $\frac{x+1}{x+5}$

Name:
Instructor:

Date:
Section:

Divide.

16. $\left(27x^3 - 18x^2 + 9x - 6\right) \div 3x$

16. ____________________

a. $9x^3 - 18x^2 + 15$

b. $-9x^2 + 9x - 6$

c. $27x^3 - 18x^2 - 3$

d. $9x^2 - 6x + 3 - \frac{2}{x}$

17. $\frac{10x^2 - 11x - 6}{2x - 3}$

17. ____________________

a. $5x + 3$

b. $5x - 4$

c. $5x + 2$

d. $5x - 6$

18. $\left(2x^3 - 5x^2 + 4x - 3\right) \div \left(x + 3\right)$

18. ____________________

a. $2x^2 - 11x + 37 - \frac{114}{x+3}$

b. $2x^2 + 14x - 22$

c. $2x^2 - 11x - 23 + \frac{21}{x+3}$

d. $2x^2 - 10x + 13 + \frac{22}{x+3}$

Name: **Date:**
Instructor: **Section:**

19. If $P(x) = 6x^3 + 4x^2 - 5x + 11$, use the remainder theorem to find $P(-2)$.

19. ____________________

a. - 31 b. – 11

c. 85 d. 65

Find the constant of variation and the variation equation.

20. y varies inversely with x; $y = 12$ when $x = -5$.

20. ____________________

a. $k = -60;\ x = \frac{-60}{y}$ b. $k = -\frac{12}{5};\ y = -\frac{12}{5}x$

c. $k = -\frac{5}{12};\ y = -\frac{5}{12}x$ d. $k = -60;\ y = \frac{-60}{x}$

Name: **Date:**
Instructor: **Section:**

Chapter 7 Rational Exponents, Radicals, and Complex Numbers
Section 7.1 Radicals and Radical Functions

Learning Objectives

1. Find square roots.
2. Approximate roots.
3. Find cube roots.
4. Find *n*th roots.
5. Find $\sqrt[n]{a^n}$ where *a* is a real number.
6. Graph square and cube root functions.

Vocabulary.
Use the choices to complete each statement.

Index	**Radical sign**	**Radicand**
Cubes	**Squares**	**True**
False		

1. In the expression $\sqrt[4]{5}$, 4 is the ____________________; 5 is the ____________________, and the $\sqrt{}$ is the ____________________.
2. True or false. The square root of – 4 is – 2. ____________________
3. The numbers 9, 25, and 81 are perfect ____________________.
4. The numbers 1, 27, and 64 are perfect ____________________.

Objective 1

Simplify.

5. $\sqrt{3600}$ 5. ____________________

6. $\sqrt{4/9}$ 6. ____________________

7. $\sqrt{0.25}$ 7. ____________________

Name: **Date:**
Instructor: **Section:**

8. $-\sqrt{49}$ 8. ____________

9. $\sqrt{-16}$ 9. ____________

10. $\sqrt{81x^{16}}$ 10. ____________

Objective 2

Use a calculator to approximate each square root to 3 decimal places.

11. $\sqrt{24}$ 11. ____________

12. $\sqrt{120}$ 12. ____________

13. $\sqrt{104}$ 13. ____________

14. $\sqrt{260}$ 14. ____________

Objective 3

Find each cube root.

15. $\sqrt[3]{64}$ 15. ____________

16. $\sqrt[3]{\dfrac{1}{-27}}$

16. ____________________

17. $\sqrt[3]{-27x^9}$

17. ____________________

18. $\sqrt[3]{1000}$

18. ____________________

Objective 4

Find each root. Assume all variables represent nonnegative real numbers.

19. $\sqrt[5]{243}$

19. ____________________

20. $\sqrt[6]{64x^{12}}$

20. ____________________

21. $\sqrt[4]{-81x^{16}}$

21. ____________________

Objective 5

Simplify, Assume all variables represent nonnegative real numbers.

22. $\sqrt{(x-5)^2}$

22. ____________________

Name:
Instructor:

Date:
Section:

23. $\sqrt[8]{y^8}$

23. ______________________

24. $\sqrt{(-5)^2}$

24. ______________________

25. $\sqrt[3]{(x-1)^3}$

25. ______________________

Objective 6

Graph the following functions.

26. $f(x) = \sqrt{x+5}$

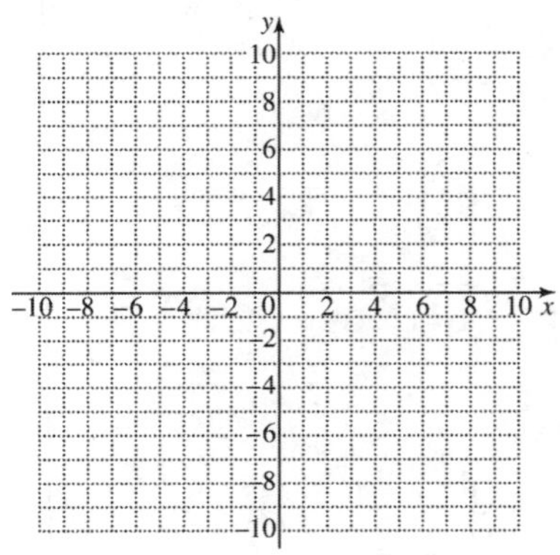

27. $g(x) = \sqrt[3]{x-2}$

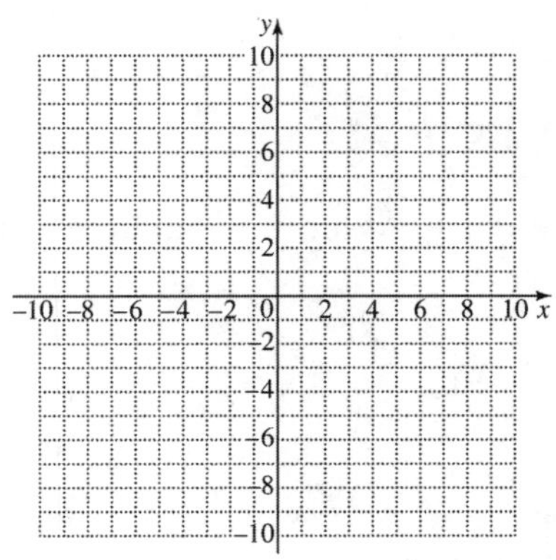

Name:
Instructor:

Date:
Section:

28. $h(x) = \sqrt{x} - 4$

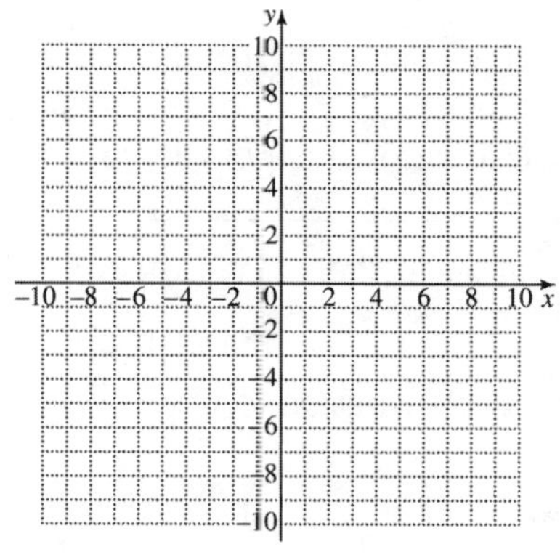

Concept Extension

Without a calculator, state what whole number the following square roots are closest to.

29. $\sqrt{389}$

29. ______________________

30. $\sqrt{2480}$

30. ______________________

Name: Date:
Instructor: Section:

Section 7.2 Rational Exponents

Learning Objectives

1. Understand the meaning of $a^{1/n}$.
2. Understand the meaning of $a^{m/n}$.
3. Understand the meaning of $a^{-m/n}$
4. Use rules for exponents to simplify expressions that contain rational exponents.
5. Use rational exponents to simplify radical expressions.

Objective 1

Use radical notation to write each expression. Simplify if possible.

1. $25^{1/2}$ 1. ____________

2. $-(32)^{1/5}$ 2. ____________

3. $(5x)^{1/3}$ 3. ____________

4. $4x^{1/4}$ 4. ____________

Objective 2

Use radical notation to write each expression. Simplify if possible.

5. $36^{3/2}$ 5. ____________

Name:
Instructor:

Date:
Section:

6. $(x+3)^{2/5}$ 6. ______________

7. $(-64)^{2/3}$ 7. ______________

8. $(4x+3)^{5/6}$ 8. ______________

Objective 3

Write with positive exponents. Simplify if possible.

9. $4^{-1/2}$ 9. ______________

10. $x^{-1/5}$ 10. ______________

11. $\dfrac{4}{5x^{-3/4}}$ 11. ______________

12. $(-32)^{-5/6}$ 12. ______________

Objective 4

Use the properties of exponents to simplify each expression. Write with positive exponents.

13. $x^{1/2}x^{1/4}$ 13. ______________

Name: **Date:**
Instructor: **Section:**

14. $\left(16y^4\right)^{3/4}$ 14. ____________

15. $\dfrac{5^{2/3}5^{1/4}}{5^{1/3}}$ 15. ____________

16. $\dfrac{\left(3x^{1/4}\right)^3}{x^{1/12}}$ 16. ____________

Multiply.

17. $\left(x^{1/2}-2\right)\left(x^{1/3}+1\right)$ 17. ____________

18. $x^{2/3}(x-2)$ 18. ____________

Factor the common factor from the given expression.

19. $x^{3/5}; x^{4/5}-3x$ 19. ____________

20. $x^{1/3}; x^{4/3}-4x^{5/3}$ 20. ____________

Name: **Date:**
Instructor: **Section:**

Objective 5

Use rational exponents to simplify each radical. Assume that all variables represent positive numbers.

21. $\sqrt[4]{y^2}$ 21. ____________

22. $\sqrt[15]{r^5 s^{10}}$ 22. ____________

23. $\sqrt[6]{x^{21} y^{15}}$ 23. ____________

Use rational exponents to write as a single expression.

24. $\sqrt[3]{x^2} \cdot \sqrt[4]{x^3}$ 24. ____________

25. $\sqrt{5} \cdot \sqrt[3]{6}$ 25. ____________

26. $\dfrac{\sqrt[3]{x^2}}{\sqrt[5]{x}}$ 26. ____________

Concept Extension

27. What times $x^{2/3}$ will the result be $x^{1/4}$?

27. ____________

28. What divided by $x^{5/8}$ will the result be $x^{7/4}$? 28. ____________

Name: **Date:**
Instructor: **Section:**

Section 7.3 Simplifying Radical Expressions

Learning Objectives

1. Use the product rule for radicals.
2. Use the quotient rule for radicals.
3. Simplify radicals.
4. Use the distance and midpoint formula.

Objective 1

Use the product rule to multiply.

1. $\sqrt{5} \cdot \sqrt{7}$ 1. ____________________

2. $\sqrt{13} \cdot \sqrt{x}$ 2. ____________________

3. $\sqrt{\frac{5}{6}} \cdot \sqrt{11}$ 3. ____________________

4. $\sqrt[5]{13ab} \cdot \sqrt[5]{2a^2b^3}$ 4. ____________________

5. $\sqrt[3]{\frac{3}{4}} \cdot \sqrt[3]{\frac{2}{5}}$ 5. ____________________

Objective 2

Use the quotient rule to simplify.

6. $\sqrt{\frac{3}{16}}$ 6. ____________________

Name: **Date:**
Instructor: **Section:**

7. $\sqrt{\frac{81}{x^2 z}}$ 7. ______________

8. $\sqrt[3]{\frac{27y}{8x^6}}$ 8. ______________

9. $\sqrt{\frac{z^{16}}{25x^{12}}}$ 9. ______________

Objective 3

Simplify.

10. $\sqrt{50}$ 10. ______________

11. $2\sqrt{8x^2}$ 11. ______________

12. $\sqrt[3]{250x^4}$ 12. ______________

13. $\sqrt[5]{-32x^{10}y}$ 13. ______________

14. $-\sqrt[4]{32x^6y^5}$ 14. ______________

Name: **Date:**
Instructor: **Section:**

Objective 4

Find the distance between each pair of points. Give an exact distance and a three-decimal-place approximation.

15. $(1,2)$ and $(-3,5)$ 15. ____________________

16. $(-3,2)$ and $(1,-3)$ 16. ____________________

17. $(0,4)$ and $(-2,-6)$ 17. ____________________

Find the midpoint of the line segment whose endpoints are given.

18. $(2,3)$ and $(-4,5)$ 18. ____________________

19. $(-3,-6)$ and $(6,4)$ 19. ____________________

20. $(-2,-1)$ and $(-8,6)$ 20. ____________________

Concept Extension

Find the midpoint and distance between the given points.

21. $(\sqrt{3},\sqrt{5})$ and $(-3\sqrt{3},6\sqrt{5})$ 21. ____________________

Name: **Date:**
Instructor: **Section:**

Section 7.4 Adding, Subtracting, and Multiplying Radical Expressions

Learning Objectives
1. Add or subtract radical expressions.
2. Multiply radical expressions.

Objective 1

Add or subtract.

1. $2\sqrt{50}-3\sqrt{125}+\sqrt{98}$ 1. ______________

2. $3\sqrt[3]{16x^3}-x\sqrt[3]{2}+\sqrt[3]{54x}$ 2. ______________

3. $3\sqrt{y^3}+\sqrt{y^5}+\sqrt{y}$ 3. ______________

4. $\sqrt[3]{\dfrac{11}{8}}-\dfrac{\sqrt[3]{11}}{6}$ 4. ______________

5. $\sqrt{\dfrac{12x}{25}} - \dfrac{\sqrt{48x}}{5}$

5. ____________________

Objective 2

Multiply, and then simplify if possible.

6. $\sqrt{3}\left(2+\sqrt{3}\right)$

6. ____________________

7. $\sqrt{6}\left(\sqrt{12}-\sqrt{24}\right)$

7. ____________________

8. $\left(2+\sqrt{3}\right)\left(4-\sqrt{3}\right)$

8. ____________________

9. $\left(2\sqrt{7}+3\sqrt{5}\right)\left(\sqrt{7}-2\sqrt{5}\right)$

9. ____________________

10. $\left(\sqrt[3]{4}-5\right)\left(3-\sqrt[3]{2}\right)$

10. ____________________

Name: **Date:**
Instructor: **Section:**

11. $\left(\sqrt{3}-2\right)^2$

11. ____________________

12. $\left(\sqrt{y+1}-3\right)^2$

12. ____________________

Concept Extension

13. Find the perimeter of a rectangle with the length of $\sqrt{108}$ feet and width of $\sqrt{75}$ feet.

13. ____________________

14. Find the length of the diagonal for the rectangle above. Round the answer to three decimal places, if necessary.

14. ____________________

Name: **Date:**
Instructor: **Section:**

Section 7.5 Rationalizing Denominators and Numerators of Radical Expressions

Learning Objectives

1. Rationalize denominators.
2. Rationalize denominators having two terms.
3. Rationalize numerators.

Vocabulary.

Use the choices to complete each statement.

Conjugate **Rationalizing the Numerator** **Rationalizing the Denominator**

$\frac{6}{6}$ $\frac{\sqrt{5}}{\sqrt{5}}$

1. ____________________ is the process of removing radicals from the numerator.

2. The ____________________ of $5+\sqrt{2}$ is $5-\sqrt{2}$.

3. To rationalize the denominator of $\frac{6}{\sqrt{5}}$, you will multiply the expression by ____________________.

Objective 1

Rationalize each denominator.

4. $\frac{4}{\sqrt{3}}$ 4. ____________________

5. $\sqrt{\frac{5}{6x}}$ 5. ____________________

6. $\frac{3}{\sqrt{8x}}$ 6. ____________________

Name: **Date:**
Instructor: **Section:**

7. $\dfrac{8x}{\sqrt[5]{8x^4}}$ 7. ______________

Objective 2

Rationalize each denominator.

8. $\dfrac{3}{2+\sqrt{x}}$ 8. ______________

9. $\dfrac{\sqrt{3}+\sqrt{5}}{4-\sqrt{2}}$ 9. ______________

10. $\dfrac{3\sqrt{x}+y}{2\sqrt{x}-\sqrt{y}}$ 10. ______________

11. $\dfrac{-7}{\sqrt{x}-3}$ 11. ______________

Name: **Date:**
Instructor: **Section:**

Objective 3

Rationalize each numerator.

12. $\sqrt{\frac{18}{5}}$

12. ____________________

13. $\frac{\sqrt{3x}}{15}$

13. ____________________

14. $\sqrt[3]{\frac{54}{243}}$

14. ____________________

15. $\frac{1+\sqrt{x}}{3}$

15. ____________________

16. $\frac{\sqrt{a}+\sqrt{b}}{\sqrt{a}-\sqrt{b}}$

16. ____________________

Name: **Date:**
Instructor: **Section:**

17. $\dfrac{\sqrt{2}-4x}{x+\sqrt{2}}$

17. ____________________

Concept Extension

18. The formula for the radius r of a sphere with a volume of V is $r=\sqrt[3]{\dfrac{3V}{4\pi}}$. Rationalize the denominator of the radical expression in this formula.

18. ____________________

Name: **Date:**
Instructor: **Section:**

Section 7.6 Radical Equations and Problem Solving

Learning Objectives
1. Solve equations that contain radical equations.
2. Use the Pythagorean Theorem to model problems.

Objective 1

Solve.

1. $\sqrt{x+1}=6$ 1. ____________________

2. $\sqrt{2x+5}+2=7$ 2. ____________________

3. $\sqrt{x-3}=-6$ 3. ____________________

4. $\sqrt[3]{x-7}=3$ 4. ____________________

Name:
Instructor:

Date:
Section:

5. $\sqrt[3]{2x-3}-2=-5$

5. ______________________

6. $\sqrt{x+8}=\sqrt{2x+6}$

6. ______________________

7. $x+\sqrt{x+1}=1$

7. ______________________

8. $\sqrt{x+2}-\sqrt{x-2}=1$

8. ______________________

9. $\sqrt{5x-1}-\sqrt{x}+2=3$

9. ______________________

Name: **Date:**
Instructor: **Section:**

Objective 2

Solve.

Find the length of the unknown side of the triangle.

10.

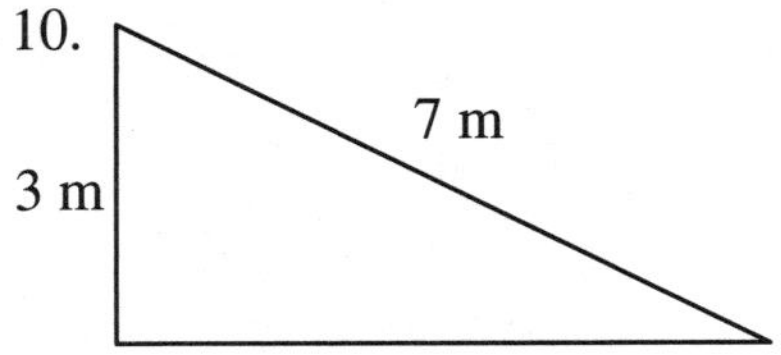

10. ______________________

11. A two cars leave Dallas, one heads north, the other heads east. The car traveling north averages a speed of 50 mph while the other car averages 45 mph. After 3 hours, how far apart are the two cars? Round your answer to the nearest whole number.

11. ______________________

12. Find the length of the unknown side.

12. ______________________

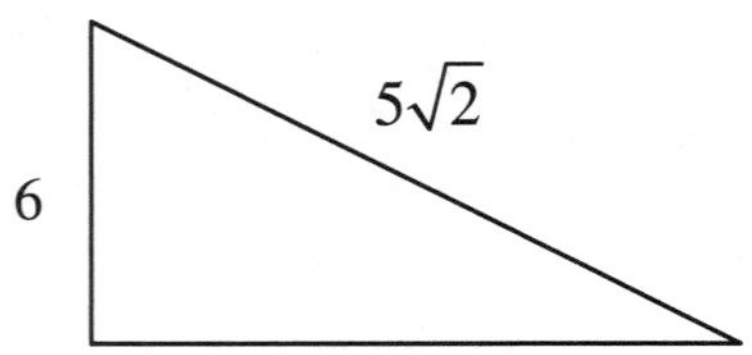

Name:
Instructor:

Date:
Section:

Concept Extension

13. $\sqrt{(x^2 - x) + 4} = 3(x^2 - x) + 2$

13. ____________________

14. Solve: $\sqrt{\sqrt{x-2} + \sqrt{x}} = \sqrt{5}$

14. ____________________

Name: **Date:**
Instructor: **Section:**

Section 7.7 Complex Numbers

Learning Objectives

1. Write square roots of negative numbers in the form bi.
2. Add or subtract complex numbers.
3. Multiply complex numbers.
4. Divide complex numbers.
5. Raise i to powers.

Vocabulary

Use the choices to complete each statement.

Complex	**Imaginary Unit**	**Pure Imaginary**
Real	**- 1**	**1**
$\sqrt{-1}$	$\sqrt{1}$	

1. In the complex number system, i is the ____________________.
2. A ____________________ number is any number that can be written in the form: $a+bi$, where both a and b are real numbers.
3. A ____________________ is the complex number, where $a = 0$ and b does not.
4. $i =$ ____________________.
5. A complex number is considered a ____________________ number, if $b = 0$.
6. $i^2 =$ ____________________.

Objective 1

Simplify.

7. $\sqrt{-36}$ 7. ____________________

8. $\sqrt{-25}$ 8. ____________________

Name: **Date:**
Instructor: **Section:**

9. $\sqrt{-5}$ 9. ________

10. $\sqrt{-72}$ 10. ________

Objective 2

Add or subtract. Write the sum or difference in the form $a+bi$.

11. $(4+i)+(3-2i)$ 11. ________

12. $(5+6i)-(4+2i)$ 12. ________

13. $(5+4i)+(3+2i)-(4-3i)$ 13. ________

14. $(4+10i)-(11+2i)$ 14. ________

Objective 3

Multiply. Write the product in the form $a+bi$.

15. $-5i \cdot 12i$ 15. ________

Name:
Instructor:

Date:
Section:

16. $4i\cdot(2+3i)$

16. ______________________

17. $(3-7i)(2-3i)$

17. ______________________

18. $(8-2i)^2$

18. ______________________

19. $(6-2i)(3+i)$

19. ______________________

Objective 4

Write each quotient in the form $a+bi$.

20. $\frac{3}{2i}$

20. ______________________

21. $\frac{4}{2-i}$

21. ______________________

Name:
Instructor:

Date:
Section:

22. $\dfrac{3+2i}{1+5i}$

22. ____________________

23. $\dfrac{3+5i}{1+i}$

23. ____________________

Objective 5

Find each power of i.

24. i^9

24. ____________________

25. i^{30}

25. ____________________

26. $(2i)^6$

26. ____________________

27. $(-4i)^3$

27. ____________________

Name:
Instructor:

Date:
Section:

Concept Extension

Write in the form $a + bi$.

28. $\dfrac{4-\sqrt{-4}}{6}$

28. ______________________

29. $\dfrac{7-\sqrt{-72}}{14}$

29. ______________________

Name: **Date:**
Instructor: **Section:**

Chapter 7 Vocabulary

Vocabulary Word	Definition	Example
Principal, or square root	$\sqrt{a}=b$ if $b^2=a$	$\sqrt{100}=10$
Cube root	$\sqrt[3]{a}=b$ if $b^3=a$	$\sqrt[3]{27}=3$
Radical Function	Function that contains a root of *x*.	$f(x)=\sqrt{x}-4$
Distance Formula	$D=\sqrt{(x_2-x_1)^2+(y_2-y_1)^2}$	$(0,2)(-2,4)$ $\sqrt{(0+2)^2+(2-4)^2}=\sqrt{4+4}=2\sqrt{2}$
Midpoint Formula	$\left(\frac{x_1+x_2}{2},\frac{y_1+y_2}{2}\right)$	$(0,2)(-2,4)$ $\left(\frac{0-2}{2},\frac{2+4}{2}\right)=\left(\frac{-2}{2},\frac{6}{2}\right)=(-1,3)$
Like Radical	Radicals with the same index and the same radicand	$5\sqrt{3}$ and $x\sqrt{3}$ are like terms
Conjugate	The conjugate of $a+b$ is $a-b$	$4c+2$ and $4c-2$ are conjugates
Rationalizing the denominator	The process of writing the denominator without the radicals.	$\frac{2}{\sqrt{3}}\bullet\frac{\sqrt{3}}{\sqrt{3}}=\frac{2\sqrt{3}}{3}$
Rationalizing the numerator	The process of writing the denominator without the radicals.	$\frac{\sqrt{2}}{5}\bullet\frac{\sqrt{2}}{\sqrt{2}}=\frac{2}{5\sqrt{2}}$
Complex number	Any number that can be written as $a+bi$	$4+3i$
Complex Conjugates	$a+bi$ and $a-bi$ are conjugates.	The conjugate of $2+3i$ is $2-3i$

Name: **Date:**
Instructor: **Section:**

Practice Test A

Raise to the power or find the root. Assume that all variables represent positive numbers. Write with only positive exponents.

1. $\sqrt{320}$ 1. ____________

2. $-\sqrt[5]{-32x^4}$ 2. ____________

3. $\left(\frac{8}{27}\right)^{-1/3}$ 3. ____________

4. $x^{1/2}\left(x^{2/3}-x^{1/4}\right)$ 4. ____________

5. $\left(a^4b^6\right)^{3/2}$ 5. ____________

6. $\left(\frac{49x^{2/3}}{y^{4/5}}\right)^{1/2}$ 6. ____________

Name:
Instructor:

Date:
Section:

7. $\sqrt{-27a^2}$

7. ____________________

8. $\sqrt{-64}$

8. ____________________

Rationalize the denominator.

9. $\sqrt{\frac{16}{5}}$

9. ____________________

10. $\frac{5+\sqrt{x}}{3-\sqrt{x}}$

10. ____________________

11. $\frac{\sqrt[3]{xy}}{\sqrt[3]{x^2y}}$

11. ____________________

12. $\frac{4+3i}{6-i}$

12. ____________________

Name:
Instructor:

Date:
Section:

Perform the indicated operation.

13. $\left(8+\sqrt{2}\right)\left(4-\sqrt{2}\right)$

13. ____________________

14. $\sqrt{5}\left(\sqrt{3}-\sqrt{2}\right)$

14. ____________________

15. $\left(\sqrt{7}-x\right)^2$

15. ____________________

16. $3\sqrt{72x^2}-x\sqrt{50}+\sqrt{8x}$

16. ____________________

Use a calculator to approximate each to three decimal places.

17. $\sqrt{172}$

17. ____________________

18. $452^{1/4}$

18. ____________________

Solve.

19. $\sqrt{2x+5}=7$

19. ____________________

Name: **Date:**
Instructor: **Section:**

20. $\sqrt{x^2+3x}=\sqrt{8x+6}$

20. ____________________

21. $x-\sqrt{x+2}=4$

21. ____________________

Perform the indicated operation and simplify.

22. $(3+4i)(5-3i)$

22. ____________________

23. $(4+i)-(3+4i)$

23. ____________________

24. Find the distance between $(-4,5)$ and $(1,-6)$.

24. ____________________

25. Find the midpoint of the line segment whose endpoints are $(-2,-4)$ and $(7,8)$.

25. ____________________

Name: **Date:**
Instructor: **Section:**

Practice Test B

Raise to the power or find the root. Assume that all variables represent positive numbers. Write with only positive exponents.

1. $\sqrt{140}$ 1. ____________

a. $2\sqrt{35}$ b. $4\sqrt{35}$

c. $5\sqrt{14}$ d. $10\sqrt{7}$

2. $-\sqrt[3]{-64y^6}$ 2. ____________

a. $8y^3$ b. $4y^2$

c. $-8y^2$ d. Not a real number

3. $\left(\frac{16}{81}\right)^{-3/4}$ 3. ____________

a. $-\frac{8}{27}$ b. $\frac{8}{27}$

c. $\frac{4}{9}$ d. $\frac{27}{8}$

4. $x^{1/4}\left(x^2 + x^{1/2}\right)$ 4. ____________

a. $x^{1/2} + x^{1/8}$ b. $x^{9/4} + x^{3/4}$

c. $x^{1/2} + x^{3/4}$ d. $x^{9/4} + x^{1/8}$

Name:
Instructor:

Date:
Section:

5. $\left(x^6 y^9\right)^{2/3}$

5. ______________

a. $x^4 y^5$

b. $x^{20/3} y^{29/3}$

c. $x^4 y^6$

d. $x^9 y^{12}$

6. $\left(\dfrac{125a^{3/4}}{b^2}\right)^{-1/3}$

6. ______________

a. $\dfrac{b^{2/3}}{5a^{1/4}}$

b. $\dfrac{5a^{1/4}}{b^{2/3}}$

c. $\dfrac{5a^{5/12}}{b^{5/3}}$

d. $-\dfrac{5a^{1/4}}{b^{2/3}}$

7. $\sqrt{-81x^6}$

7. ______________

a. $9x^3 i$

b. $-9x^3$

c. $-9x^3$

d. $9x^4 i$

8. $\sqrt{-125}$

8. ______________

a. $-5\sqrt{5}$

b. $5\sqrt{5}i$

c. $-5i$

d. $25i$

Name: **Date:**
Instructor: **Section:**

Rationalize the denominator.

9. $\sqrt{\dfrac{25}{6}}$

9. ____________________

a. $\dfrac{5\sqrt{6}}{6}$ b. $\dfrac{5}{6}$

c. $\dfrac{5\sqrt{3}}{2}$ d. $\dfrac{5}{2}$

10. $\dfrac{6-\sqrt{3}}{5+\sqrt{2}}$

10. ____________________

a. $\dfrac{\sqrt{6}+5\sqrt{3}-6\sqrt{2}-30}{29}$ b. $\dfrac{\sqrt{6}+5\sqrt{3}-6\sqrt{2}-30}{23}$

c. $\dfrac{\sqrt{6}-5\sqrt{3}-6\sqrt{2}+30}{29}$ d. $\dfrac{\sqrt{6}-5\sqrt{3}-6\sqrt{2}+30}{23}$

11. $\dfrac{\sqrt[4]{ab^2}}{\sqrt[4]{a^2b}}$

11. ____________________

a. $\dfrac{\sqrt[4]{a^3b}}{a}$ b. $\dfrac{\sqrt[4]{ab^3}}{ab}$

c. $\dfrac{\sqrt[4]{a^3b^2}}{a}$ d. $\dfrac{\sqrt[4]{a^3b^3}}{ab}$

Name: **Date:**
Instructor: **Section:**

12. $\dfrac{5-2i}{8+2i}$ 12. ______________

a. $\dfrac{5}{8}-i$ b. $\dfrac{5}{4}-\dfrac{1}{2}i$

c. $\dfrac{9}{17}-\dfrac{13}{34}i$ d. $\dfrac{11}{17}-\dfrac{3}{24}i$

Perform the indicated operation.

13. $(7-\sqrt{3})(8-\sqrt{3})$ 13. ______________

a. $59-15\sqrt{3}$ b. $53-\sqrt{3}$

c. $53+\sqrt{3}$ d. $15\sqrt{3}+59$

14. $\sqrt{10}(\sqrt{12}+\sqrt{15})$ 14. ______________

a. $4\sqrt{30}+25\sqrt{6}$ b. $2\sqrt{30}+5\sqrt{6}$

c. $2\sqrt{30}-5\sqrt{6}$ d. $\sqrt{120}+\sqrt{50}$

15. $(\sqrt{2x}-5)^2$ 15. ______________

a. $2x-10\sqrt{2x}+25$ b. $4x+10\sqrt{2x}-25$

c. $27-5\sqrt{2x}$ d. $27x-10\sqrt{2x}$

Name: **Date:**
Instructor: **Section:**

16. $5\sqrt{108} + x\sqrt{75} - \sqrt{27}$ 16. ____________

a. $32x\sqrt{3}$ b. $27\sqrt{3} + 5x$

c. $5x\sqrt{3} + 3$ d. $(5x + 27)\sqrt{3}$

Use a calculator to approximate each to three decimal places.

17. $\sqrt{642}$ 17. ____________

a. 25.338 b. 24.447

c. 25.663 d. 26.116

18. $225^{1/3}$ 18. ____________

a. 6.893 b. 7.042

c. 6.082 d. 6.889

Solve.

19. $\sqrt{4x - 7} = 5$ 19. ____________

a. 8 b. 3.5

c. 3 d. $\varnothing$

20. $\sqrt{2x + 16} = \sqrt{5x + 7}$ 20. ____________

a. 4 b. -3

c. 3 d. $\varnothing$

Name:
Instructor:

Date:
Section:

21. $2x + \sqrt{x-1} = 3$

21. ____________________

a. $\frac{5}{4}$

b. 1

c. 2

d. $\varnothing$

Perform the indicated operation and simplify.

22. $(4-5i)(2+6i)$

22. ____________________

a. $-22+34i$

b. $38-14i$

c. $-22-34i$

d. $38+14i$

23. $(6+2i)-(7+5i)$

23. ____________________

a. $-1+7i$

b. $-1-3i$

c. $13-3i$

d. $13+7i$

24. Find the distance between $(-2,-6)$ and $(4,8)$.

24. ____________________

a. $2\sqrt{2}$

b. $4\sqrt{29}$

c. $2\sqrt{58}$

d. $4\sqrt{10}$

Name: **Date:**
Instructor: **Section:**

25. Find the midpoint of the line segment whose endpoints are $(1,7)$ and $(-4,-8)$.

25. ____________________

a. $\left(\frac{7}{2},\frac{11}{2}\right)$ b. $\left(\frac{5}{2},\frac{15}{2}\right)$

c. $(3,7)$ d. $\left(-\frac{3}{2},-\frac{1}{2}\right)$

Name: **Date:**
Instructor: **Section:**

Chapter 8 Quadratic Equations and Functions
Section 8.1 Solving Quadratic Equations by Completing the Square

Learning Objective
1. Use the square root property to solve quadratic equations.
2. Solve quadratic equations by completing the square.
3. Use quadratic equations to solve problems.

Vocabulary
Use the choices to complete each statement.

Completing the square	**Quadratic Equation**	$\sqrt{b}$
$-\sqrt{b}$	$\pm\sqrt{b}$	b^2
$\frac{b}{2}$	$\left(\frac{b}{2}\right)^2$	

1. A ________________________ is an equation that can be written in the form $ax^2 + bx + c = 0,\ a \neq 0$.
2. According to the square root property, if $a^2 = b$ then $a =$ __________________, if b is a real number.
3. One method to solve a quadratic equation is ____________________________, this method is the process of rewriting one side to be a perfect square trinomial.
4. To solve a quadratic equation by completing the square, one will add _________________ to both sides.

Objective 1

Use the square root property to solve each equation.

5. $x^2 = 36$ 5. ______________________

6. $(x+7)^2 = 16$ 6. ______________________

7. $x^2 + 4 = 9$ 7. ______________________

Name: **Date:**
Instructor: **Section:**

8. $(x+2)^2 = 10$ — 8. ____________________

9. $(x-3)^2 = -9$ — 9. ____________________

10. $(x-4)^2 = -12$ — 10. ____________________

Objective 2

Solve each equation by completing the square.

11. $x^2 + 6x + 2 = 0$ — 11. ____________________

12. $x^2 + 8x - 6 = 0$ — 12. ____________________

13. $x^2 + 3x + 5 = 0$ — 13. ____________________

Name: **Date:**
Instructor: **Section:**

14. $3y^2+6y-4=0$

14. ____________________

15. $x^2+6x+12=0$

15. ____________________

16. $x^2+4x+8=0$

16. ____________________

17. $2x^2+6x+11=0$

17. ____________________

Objective 3

Use the formula $A=P(1+r)^t$ to solve.

18. Find the rate r at which \$3000 compounded annually grows to \$4320 in 2 years.

18. ____________________

19. Find the rate r at which \$4200 compounded annually grows to \$8232 in 2 years.

19. ____________________

Name: **Date:**
Instructor: **Section:**

Concept Extension

20. An right triangle has legs of equal length. If the hypotenuse is 48 centimeters long, find the length of each leg.

20. ______________________

21. The area of a square is 36 square centimeters. Find the length of the diagonal of the square.

21. ______________________

Name: **Date:**
Instructor: **Section:**

Section 8.2 Solving Quadratic Equations by the Quadratic Formula

Learning Objectives

1. Solve quadratic equations by using the quadratic formula.
2. Determine the number and type of solutions of a quadratic equation by using the discriminant.
3. Solve geometric problems modeled by quadratic equations.

Objective 1

Use the quadratic formula to solve each equation.

1. $x^2+16x+15=0$ 1. ____________________

2. $x^2+14x+45=0$ 2. ____________________

3. $2x^2+3x-6=0$ 3. ____________________

4. $3x^2+12x-20=0$ 4. ____________________

5. $-2x^2+x-5=0$ 5. ____________________

Name: **Date:**
Instructor: **Section:**

6. $\frac{2}{5}y^2 + \frac{1}{5}y + \frac{3}{5} = 0$ 6. ____________________

Use the discriminant to determine the number and types of solutions of each equation.

7. $2x^2 + 4x - 5 = 0$ 7. ____________________

8. $x^2 + 3x + 7 = 0$ 8. ____________________

9. $6 = 4x - 5x^2$ 9. ____________________

10. $15x^2 + 13x - 27 = 0$ 10. ____________________

Name: **Date:**
Instructor: **Section:**

Objective 3

Solve.

11. Nancy, Thelma, and John Varner live on a corner lot. Often, neighborhood children across their lot to save walking distance. Give the diagram below, approximate to the nearest foot how many feet of walking distance is saved from cutting across their property instead of walking around the lot.

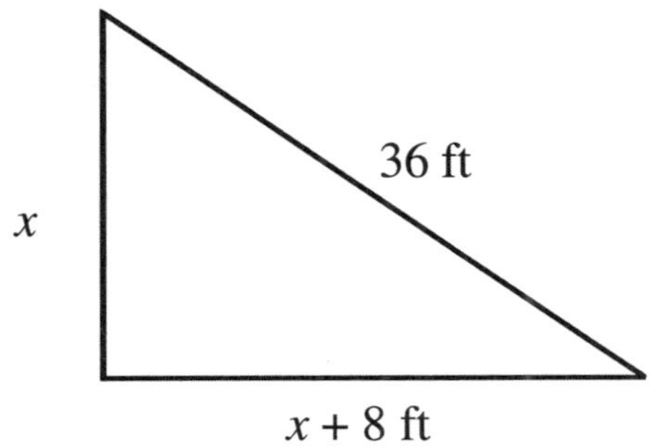

11. ______________________

12. A rectangle is 6 more meters long as it is high. If the area of the rectangle is 22 square meters. Find dimensions of the rectangle.

12. ______________________

Concept Extension

Use the quadratic equation to solve each quadratic equation.

13. $x^2+\sqrt{5}x-3=0$

13. ______________________

14. $3x^2-\sqrt{6}x+2=0$

14. ______________________

Name: **Date:**
Instructor: **Section:**

Section 8.3 Solving Equations by Using Quadratic Methods

Learning Objectives
1. Solve various equations that are quadratic in form.
2. Solve problems that lead to quadratic equations.

Objective 1

Solve.

1. $x = \sqrt{6x-5}$ 1. ____________________

2. $\sqrt{17-x} = (x+3)$ 2. ____________________

3. $2x-3 = \sqrt{3x}$ 3. ____________________

4. $\frac{2}{x} + \frac{3}{x-1} = 1$ 4. ____________________

Name:
Instructor:

Date:
Section:

5. $\frac{2}{x^2}=\frac{4}{x+1}$

5. ____________________

6. $\frac{4}{x+3}-\frac{2}{x-3}=2$

6. ____________________

7. $\frac{5}{x^2+7x+12}=\frac{3}{x+4}-\frac{6}{x+3}$

7. ____________________

8. $x^4-64=0$

8. ____________________

9. $x^4+3x^2-4=0$

9. ____________________

Name: **Date:**
Instructor: **Section:**

10. $x^{2/3} - 3x^{1/3} - 10 = 0$

10. ______________________

11. $x + 2x^{1/2} - 5 = 0$

11. ______________________

12. $(x-1)^2 - 3(x-1) - 6 = 0$

12. ______________________

Objective 2

Solve.

13. A jogger ran 3 miles, decreased her speed by 1 mile per hour and then ran another 4 miles. If her total time jogging was $1\frac{3}{5}$ hours, find her speed for each part of her run.

13. ______________________

Name:
Instructor:

Date:
Section:

14. The product of a number and 11 more than that number is 80. Find the number.

14. ____________________

15. Joe and Sarah can complete a job if they work together in 8 hours. If it takes Sarah 2 more hours than Joe to do it alone, find how long it will take each of them to complete the job separately to the nearest tenth of an hour.

15. ____________________

Concept Extension

Solve.

16. $x^3 - 125 = 0$

16. ____________________

17. $x^{-2} + 6x^{-1} + 5 = 0$

17. ____________________

Name: **Date:**
Instructor: **Section:**

Name:
Instructor:

Date:
Section:

Section 8.4 Nonlinear Inequalities in One Variable

Learning Objectives
1. Solve polynomial inequalities of degree 2 or greater.
2. Solve inequalities that contain rational expressions with variables in the denominator.

Objective 1

Solve each quadratic inequality. Write the solution in interval notation.

1. $(x-3)(x+4) \le 0$ 1. ____________________

2. $(x-5)(x+1) > 0$ 2. ____________________

3. $x^2 + 4x - 12 < 0$ 3. ____________________

4. $(x+4)(x-3)(x+1) \ge 0$ 4. ____________________

Name: **Date:**
Instructor: **Section:**

5. $(x^2-25)(x^2-1)>0$ 5. ____________

6. $(2x-3)(3x+5)(x-4)\le 0$ 6. ____________

Objective 2

Solve each inequality. Write the solution set in interval notation.

7. $\dfrac{2}{x+3}<6$ 7. ____________

8. $\dfrac{x+3}{x-4}\ge 0$ 8. ____________

Name: **Date:**
Instructor: **Section:**

9. $\dfrac{x^2}{14-5x} < 1$

9. ____________________

10. $\dfrac{x(x-3)}{(x+2)(x-5)} \geq 0$

10. ____________________

Concept Extension

11. Explain why $(x+5)(x-4) > 0$ and $\dfrac{x+5}{x-4} > 0$ have the same solution.

11. ____________________

12. Find all numbers that satisfy the following sentence:
A number minus the product of 4 and the number's reciprocal is positive.

12. ____________________

Name:
Instructor:

Date:
Section:

Section 8.5 Quadratic Functions and Their Graphs

Learning Objectives

1. Graph quadratic functions of the form $f(x)=x^2+k$.
2. Graph quadratic functions of the form $f(x)=(x-h)^2$.
3. Graph quadratic functions of the form $f(x)=(x-h)^2+k$.
4. Graph quadratic functions of the form $f(x)=ax^2$.
5. Graph quadratic functions of the form $f(x)=a(x-h)^2+k$

Objective 1

Sketch the graph of each quadratic function. Label the vertex, and sketch and label the axis of symmetry.

1. $f(x)=x^2+2$

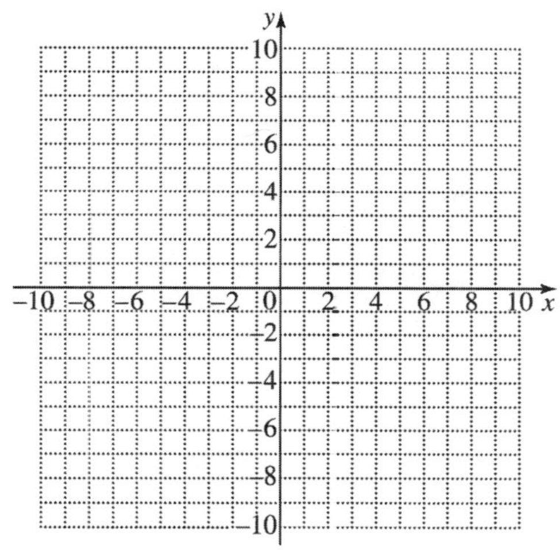

2. $f(x)=x^2-4$

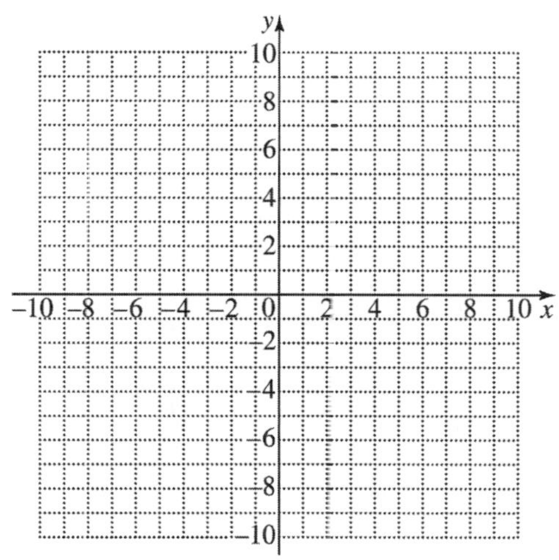

Name:
Instructor:

Date:
Section:

3. $f(x) = x^2 + 6$

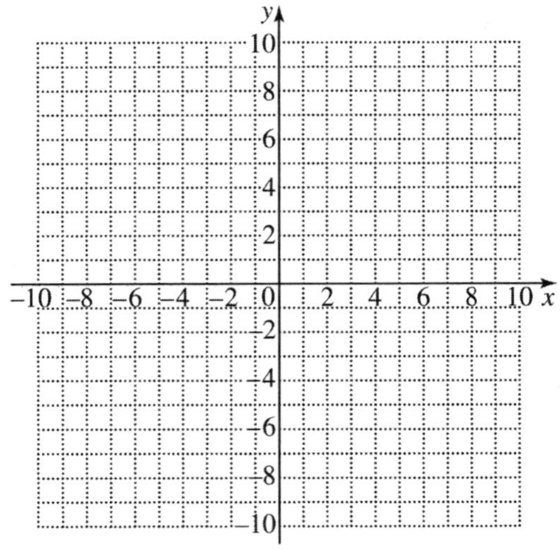

Objective 2

Sketch the graph of each quadratic equation. Label the vertex, and sketch and label the axis of symmetry.

4. $h(x) = (x+2)^2$

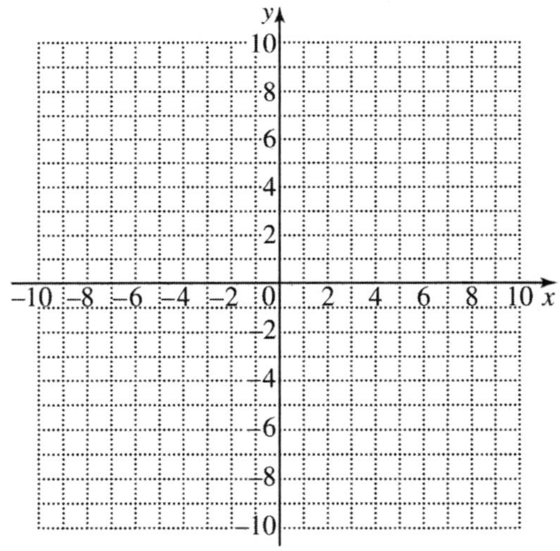

Name:
Instructor:

Date:
Section:

5. $f(x) = (x-4)^2$

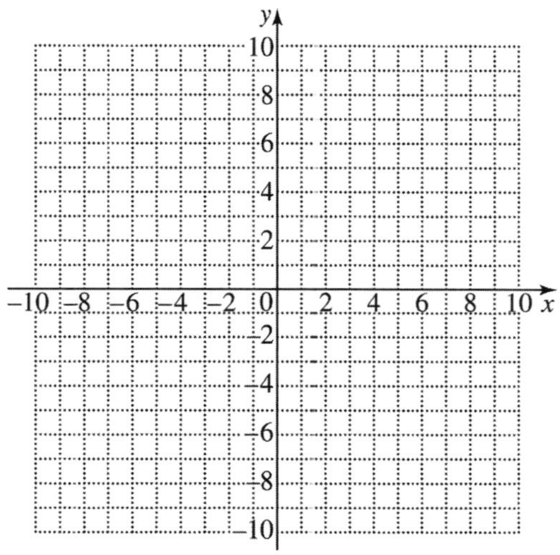

6. $f(x) = (x+3)^2$

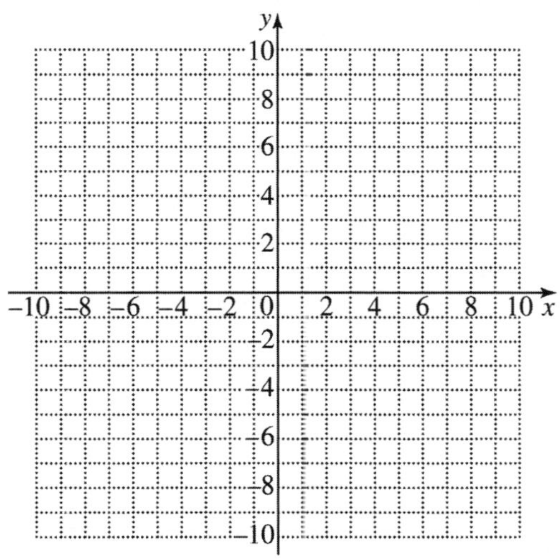

Name: **Date:**
Instructor: **Section:**

Objective 3

Sketch the graph of each quadratic equation. Label the vertex, and sketch and label the axis of symmetry.

7. $f(x) = (x-2)^2 + 5$

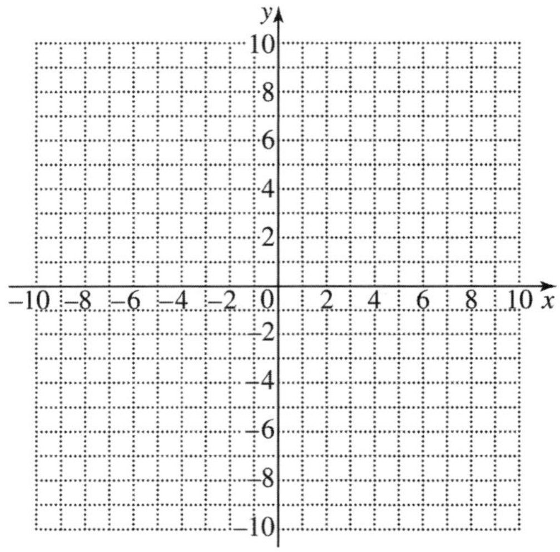

8. $f(x) = (x+1)^2 - 4$

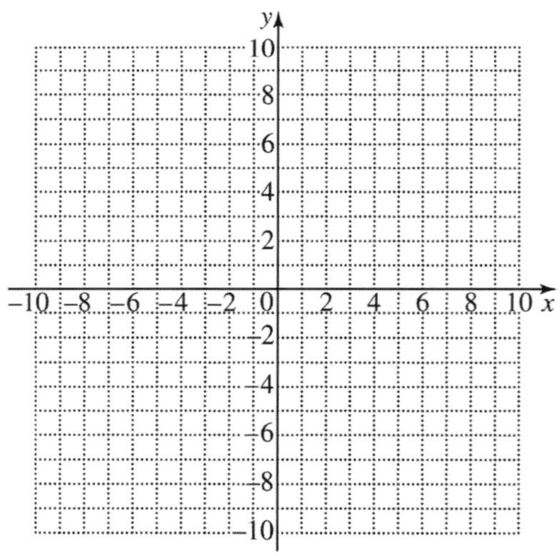

Name:
Instructor:

Date:
Section:

9. $f(x) = (x-3)^2 - 2$

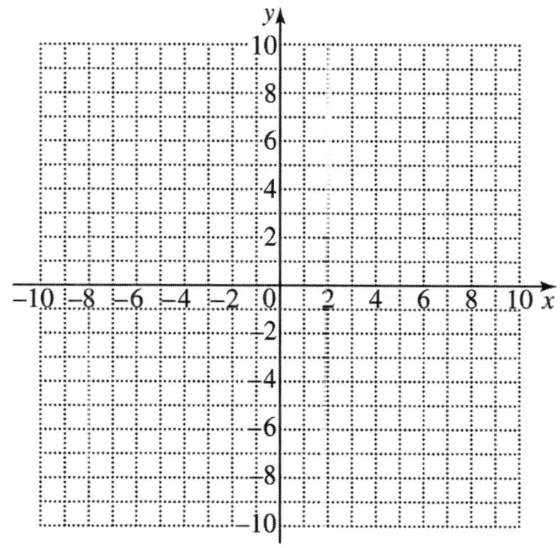

Objective 4

Sketch the graph of each quadratic equation. Label the vertex, and sketch and label the axis of symmetry.

10. $f(x) = 3x^2$

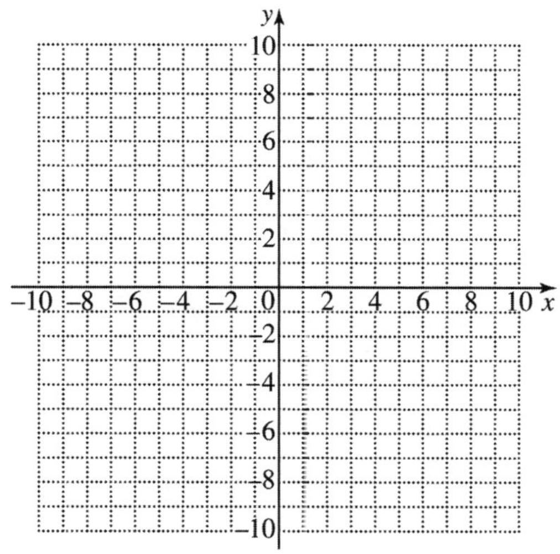

Name: **Date:**
Instructor: **Section:**

11. $f(x) = -4x^2$

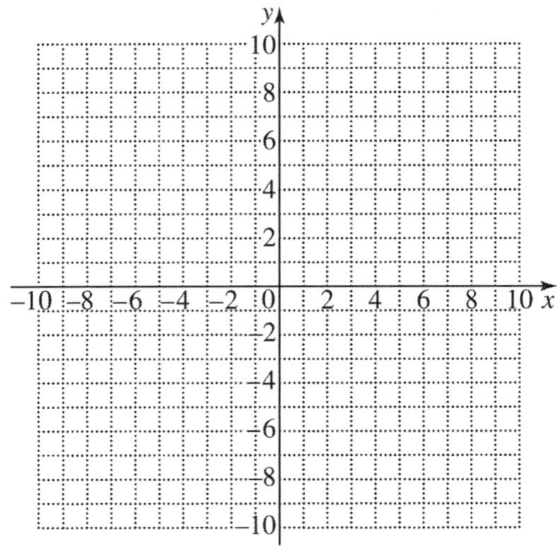

12. $f(x) = -\frac{1}{2}x^2$

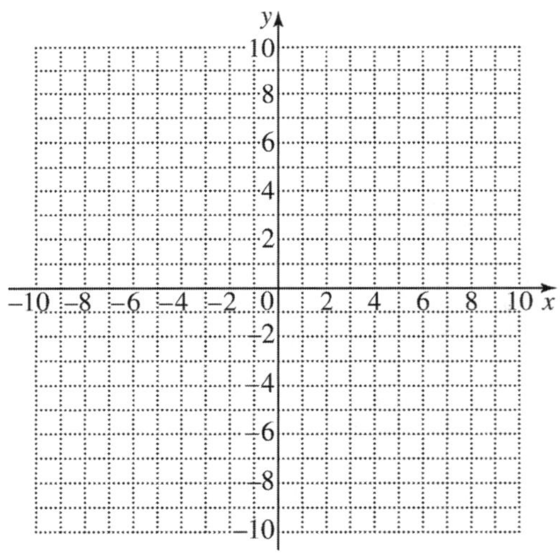

Name:
Instructor:

Date:
Section:

13. $f(x)=\frac{1}{4}x^2$

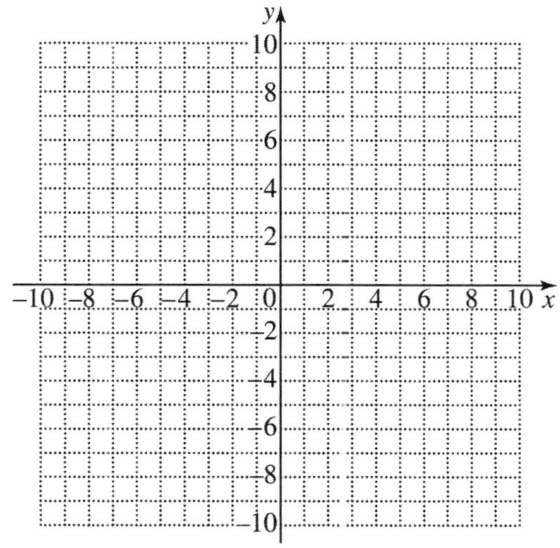

Objective 5

Sketch the graph of each quadratic equation. Label the vertex, and sketch and label the axis of symmetry.

14. $h(x)=-3(x+3)^2+1$

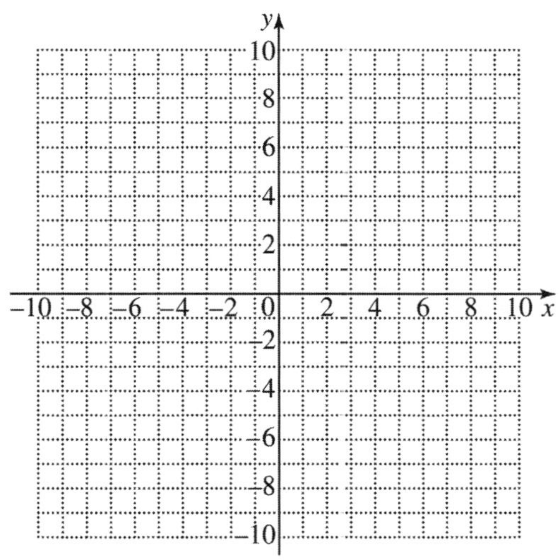

Name: **Date:**
Instructor: **Section:**

15. $f(x) = 2(x-1)^2 + 2$

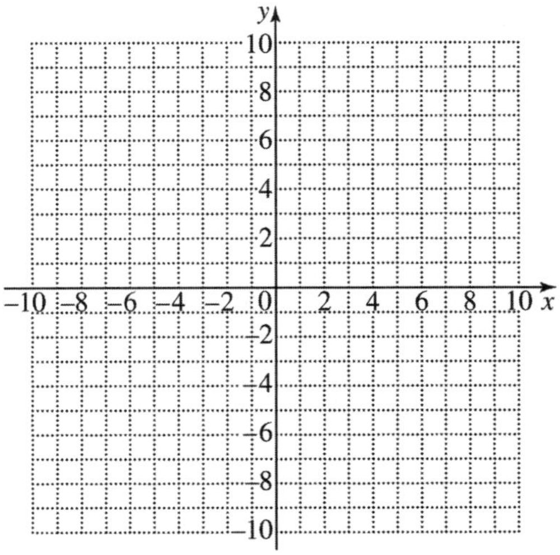

16. $f(x) = \frac{1}{2}(x+2)^2 + 3$

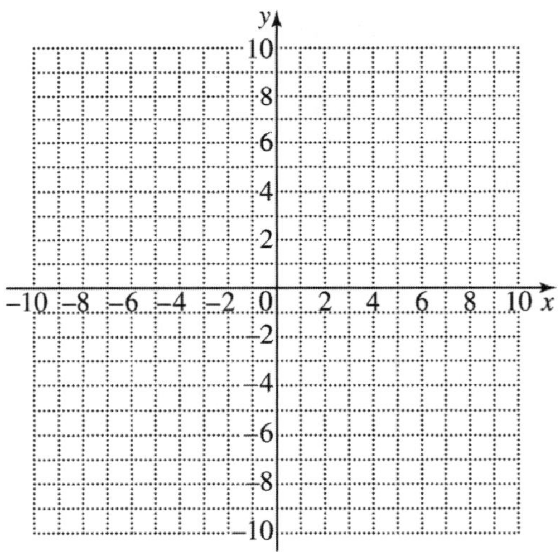

Name:
Instructor:

Date:
Section:

Concept Extension

17. Write the equation of the parabola that has the same shape of $f(x) = -2x^2$ but with the vertex of $(4,-6)$.

17. ____________________

18. What about $(-7,9)$?

18. ____________________

Name:
Instructor:

Date:
Section:

Section 8.6 Further Graphing of Quadratic Functions

Learning Objectives

1. Write quadratic functions in the form $y = a(x-h)^2 + k$.
2. Derive the formula for finding the vertex of a parabola.
3. Find the minimum or maximum value of a quadratic function.

Objective 1

Rewrite the following functions in the form $y = a(x-h)^2 + k$. State the vertex of each function.

1. $f(x) = x^2 - 2x - 15$

1. ________________________

2. $f(x) = x^2 + 6x + 17$

2. ________________________

3. $f(x) = x^2 + 5x + 18$

3. ________________________

4. $f(x) = 2x^2 + 6x - 11$

4. ________________________

5. $f(x) = -3x^2 + 12x + 8$

5. ____________________

6. $f(x) = -x^2 + 8x - 9$

6. ____________________

Objective 2

7. Take the equation $y = ax^2 + bx + c$ and using the variables a, b, and c to rewrite the equation to be in the form $y = a(x-h)^2 + k$. What value did you get in place of the h?

7. ____________________

8. The value you created from # 7 is the x-value of your vertex, how would you get the y-value of the vertex?

8. ____________________

Name: **Date:**
Instructor: **Section:**

Using the formula you found in # 7 and # 8, find the vertex of the following functions.

9. $y = x^2 + 4x + 7$ 9. ____________________

10. $f(x) = 2x^2 + 6x + 9$ 10. ____________________

11. $f(x) = 3x^2 - 6x - 13$ 11. ____________________

12. $f(x) = -x^2 + 4x - 5$ 12. ____________________

Objective 3

Solve.

13. If a projectile is fired straight upward from the ground with an initial speed of 96 feet per second, then its height h in feet after t seconds is given by the equation $h(t) = -16t^2 + 96t$. Find the maximum height of the projectile.

13. ____________________

Name: **Date:**
Instructor: **Section:**

14. The local high school bowling team is selling candles to raise money for their scholarship fund. The profit in dollars is $P(x) = 420x - x^2$.

a. How many candles must be sold to maximize profit.

14a. ______________________

b. Find the maximum profit.

14b. ______________________

Concept Extension

Find the maximum or minimum value of each function. Approximate to two decimal points. State whether it is a maximum or minimum value.

15. $f(x) = 1.2x^2 + 3.6x + 4.5$

15. ______________________

16. $f(x) = -0.3x^2 + 1.6x - 1.2$

16. ______________________

Name: **Date:**
Instructor: **Section:**

Chapter 8 Vocabulary

Vocabulary Word	Definition	Example
Square root property	If $a^2 = b$ then $a = \pm\sqrt{b}$	$\sqrt{36} = \pm 6$
Quadratic Formula	$x = \frac{-b \pm \sqrt{b^2 - 4ac}}{2a}$	
Quadratic function	$y = a(x-h)^2 + k$ Vertex: (h,k) if a is positive parabola opens upward if a is negative parabola opens downward.	$y = 4(x-3)^2 + 5$ Vertex: $(3,5)$, parabola opens upward since $a = 4$
Quadratic equation	$y = ax^2 + bx + c$ Vertex $\left(-\frac{b}{2a}, f\left(-\frac{b}{2a}\right)\right)$	$y = 2x^2 - 8x + 9$ Vertex at $(2,1)$

Name:
Instructor:

Date:
Section:

Practice Test A

Solve each equation for the variable.

1. $x^2 - 3x - 5 = 0$

1. ____________________

2. $35x^2 + 46x - 16 = 0$

2. ____________________

3. $x^2 + 6x = -3$

3. ____________________

4. $(y-7)^2 = 36$

4. ____________________

5. $a^4 = 8a^2 + 9$

5. ____________________

Name: **Date:**
Instructor: **Section:**

6. $(x-2)^2-16(x-2)+63=0$

6. ____________________

7. $x^3+2x^2=9x+18$

7. ____________________

8. $\dfrac{x}{x+4}-\dfrac{2}{x-4}=\dfrac{3}{x^2-16}$

8. ____________________

Solve the following functions by completing the square.

9. $x^2-12x+19=0$

9. ____________________

10. $4c^2-24c+15=0$

10. ____________________

Name: **Date:**
Instructor: **Section:**

Solve each inequality for x. Write the solution set in interval notation.

11. $(3x+5)(x-8) \geq 0$

11. ____________________

12. $\frac{(x+7)}{(2x-9)} < 0$

12. ____________________

13. $2x^2 - 3 > -x$

13. ____________________

Name:
Instructor:

Date:
Section:

Graph each function. Label the vertex.

14. $f(x) = (x+3)^2 + 6$

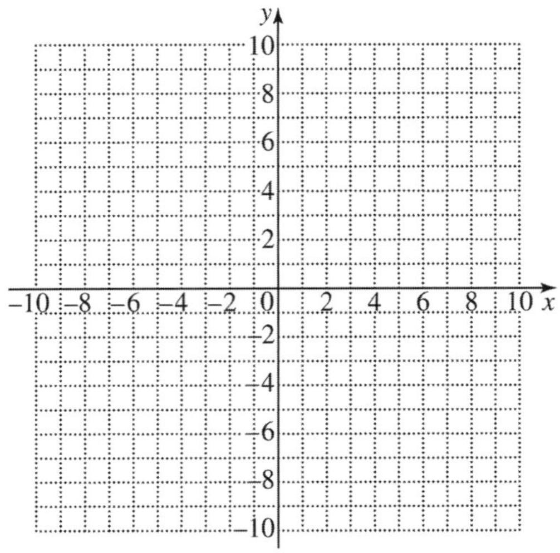

15. $h(x) = -5x^2 - 4$

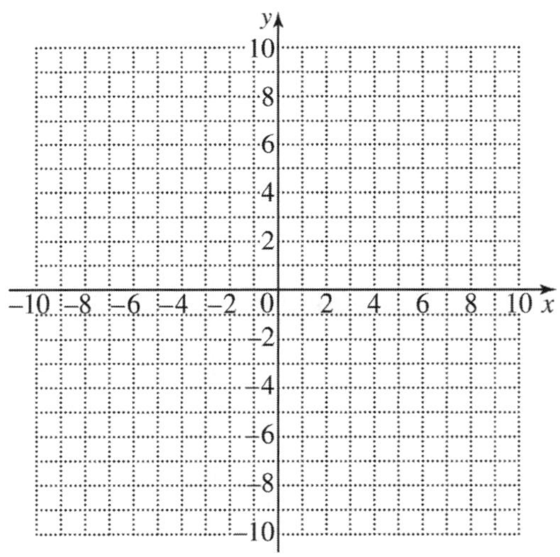

Name:
Instructor:

Date:
Section:

16. $g(x) = x^2 + 4x + 9$

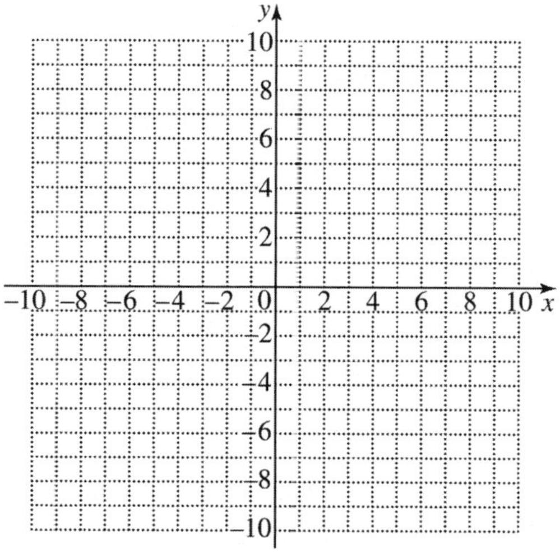

17. $f(x) = -2x^2 + 16x - 25$

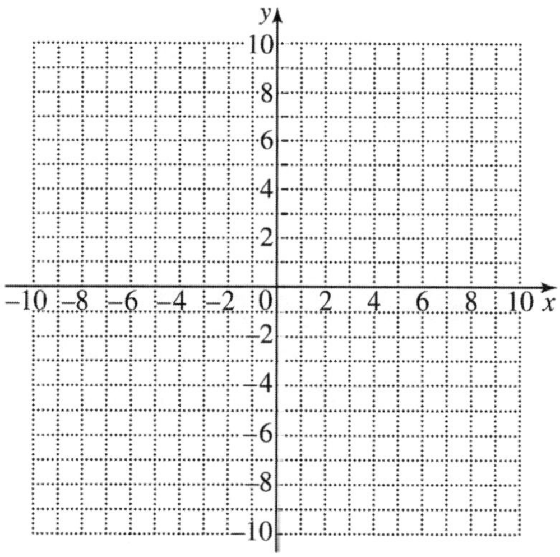

Name: **Date:**
Instructor: **Section:**

Solve.

18. One number is eight more than another number, if the product of these two number is 273, what are the two numbers.

18. ____________________

19. A triangle has a height of $x-8$ and a base of $2x-8$. Given that the area of the triangle is 32 feet, find the base and the height of the triangle.

19. ____________________

20. The height an object follows when it is shot out of a cannon follows the path of a parabola. The equation for height is given to be $h(t)=-16t^2+144t$, where t is time in second and h is height in feet. At what time will the object be at its maximum height?

20. ____________________

Name:
Instructor:

Date:
Section:

Practice Test B

Solve each equation for the variable.

1. $3x^2 + 20 = 19x$ 1. ________________

a. $\frac{4}{3}$ and 5

b. 4 and 5

c. 20 and $\frac{1}{3}$

d. $\frac{20}{3}$ and 1

2. $(x+7)^2 = 18$ 2. ________________

a. -1 and - 4

b. 11 and – 25

c. $7 \pm 2\sqrt{3}$

d. $-7 \pm 3\sqrt{2}$

3. $r^2 - 3r - 54 = 0$ 3. ________________

a. – 9 and - 6

b. 9 and – 6

c. 9 and 6

d. – 9 and 6

4. $x^2 + 3x + 7 = 0$ 4. ________________

a. $\dfrac{-3 \pm \sqrt{19}i}{2}$

b. $\dfrac{-3 \pm \sqrt{37}}{2}$

c. 7 and 1

d. $\dfrac{-2 \pm \sqrt{33}i}{6}$

5. $\dfrac{x}{x-3} - \dfrac{5}{x-4} = \dfrac{-3}{x^2 - 7x + 12}$ 5. ________________

a. 6

b. 6 and 3

c. 3

d. $\varnothing$

Name: **Date:**
Instructor: **Section:**

6. $6(x+7)^2 - 5(x+7) - 4 = 0$ 6. ____________________

a. $-\frac{1}{2}$ and $\frac{4}{3}$ b. -7 and -4

c. $-\frac{15}{2}$ and $-\frac{17}{3}$ d. 7 and 4

7. $w^4 + 8w^2 + 12 = 0$ 7. ____________________

a. $\pm\sqrt{6}$ and $\pm\sqrt{2}$ b. 6 and 2

c. 4 and 3 d. $\pm\sqrt{6}i$ and $\pm\sqrt{2}i$

8. $2x^3 + 3x^2 - 8x - 12 = 0$ 8. ____________________

a. 6, -1, and 2 b. ± 2 and $-\frac{3}{2}$

c. 3, 4, and – 1 d. 4, 3, and – 2

Solve the following functions by completing the square.

9. $x^2 + 10x - 13 = 0$ 9. ____________________

a. $-5 \pm 2\sqrt{3}$ b. –3 and 13

c. $-5 \pm \sqrt{38}$ d. –13 and 1

Name: **Date:**
Instructor: **Section:**

10. $-3c^2+36c-18=0$ 10. ____________

a. $-6\pm\sqrt{30}$ b. -6 and 3

c. -4 and 14 d. 6 and 1

Solve each inequality for x. Write the solution set in interval notation.

11. $(x-2)(4x+5)<0$ 11. ____________

a. $\left(-\frac{5}{4},2\right)$ b. $\left(-\infty,-\frac{5}{4}\right)\cup(2,\infty)$

c. $\left[-2,\frac{5}{4}\right]$ d. $(-\infty,-2]\cup\left[\frac{5}{4},\infty\right)$

12. $\dfrac{(6x+4)}{(x-3)}\geq 0$ 12. ____________

a. $\left[-\frac{2}{3},3\right)$ b. $(-\infty-3)\cup\left(\frac{2}{3},\infty\right)$

c. $\left(-\infty,-\frac{2}{3}\right]\cup(3,\infty)$ d. $\left(-3,\frac{2}{3}\right)$

13. $x^2+5x<6$ 13. ____________

a. $(-6,1)$ b. $(-\infty-1)\cup(6,\infty)$

c. $(-\infty,6)$ d. $(-1,6)$

Name: **Date:**
Instructor: **Section:**

Find the vertex of the following quadratic equations.

14. $f(x) = 5(x-4)^2 - 22$ 14. ____________

a. $(5,22)$ b. $(-20,-22)$

c. $(4,-22)$ d. $(4,22)$

15. $g(x) = 16x^2 + 120$ 15. ____________

a. $(0,-120)$ b. $(0,120)$

c. $(-16,120)$ d. $(16,120)$

Find the axis of symmetry of the following quadratic equation.

16. $h(x) = x^2 - 6x - 12$ 16. ____________

a. $x = -12$ b. $x = -6$

c. $x = -3$ d. $x = 3$

Find the y-intercept of the following quadratic equation.

17. $g(x) = -2(x-5)^2 - 11$ 17. ____________

a. $(0,-61)$ b. $(0,-5)$

c. $(0,-2)$ d. $(0,-11)$

Name:
Instructor:

Date:
Section:

Solve.

18. Find two consecutive even integers whose product is 840.

18. ____________________

a. 42 and 20 b. 28 and 30

c. 21 and 40 d. 35 and 24

19. The cost function for a company to make dolls is $C(x) = 4x^2 - 240x + 12000$, where x is the number of dolls sold. Find the number of dolls that need to be made to minimize costs.

19. ____________________

a. 45 dolls b. 40 dolls

c. 30 dolls d. 20 dolls

20. The quotient of five and eight more than four times a number, yields the quotient of the number and the sum of the number and two. Find the number.

20. ____________________

a. -2 or $\frac{5}{4}$ b. -2 or 5

c. 2 or 5 d. -5 or $\frac{1}{2}$

Name: **Date:**
Instructor: **Section:**

Chapter 9 Exponential and Logarithmic Functions
Section 9.1 The Algebra of Functions; Composite Functions

Learning Objectives

1. Add, subtract, multiply, and divide functions.
2. Construct composite functions.

Objective 1

For the functions *f* and *g*, find a. $(f+g)(x)$, b. $(f-g)(x)$, c. $(f \cdot g)(x)$, and d. $\left(\frac{f}{g}\right)(x)$.

1. $f(x)=x-5,\ \ g(x)=2x+3$

1a. ____________

1b. ____________

1c. ____________

1d. ____________

2. $f(x)=x^2+1,\ \ g(x)=5x$

2a. ____________

2b. ____________

2c. ____________

2d. ____________

Name: **Date:**
Instructor: **Section:**

3. $f(x)=\sqrt{x+5},\ \ g(x)=4x^2$

3a. ____________________

3b. ____________________

3c. ____________________

3d. ____________________

4. $f(x)=8x+3,\ \ g(x)=\sqrt[3]{x}$

4a. ____________________

4b. ____________________

4c. ____________________

4d. ____________________

Objective 2

If $f(x)=x^2+3x-4$, $g(x)=2x-5$ and $h(x)=\sqrt[3]{x}$ find the following compositions.

5. $(f \circ g)(4)$

5. ____________________

6. $(g \circ h)(-64)$

6. ____________________

7. $(g \circ f)(-6)$

7. ____________________

Name: **Date:**
Instructor: **Section:**

Find $(f \circ g)(x)$ and $(g \circ f)(x)$.

8. $f(x) = x^2 + 1,\ g(x) = 5x$

8. ______________________

9. $f(x) = x - 7,\ \ g(x) = \sqrt{x+3}$

9. ______________________

10. $f(x) = 3x^2 - 4x + 5,\ g(x) = \sqrt{x}$

10. ______________________

11. $f(x) = x^2 - 9,\ g(x) = \sqrt[3]{2x+5}$

11. ______________________

If $f(x) = 3x$, $g(x) = \sqrt{x}$, and $h(x) = x^2 + 2$, write each function as a composition using two of the given functions.

12. $H(x) = \sqrt{x^2 + 2}$

12. ______________________

Name:
Instructor:

Date:
Section:

Concept Extension

Find the following given these values:

$f(-1)=6$	$g(-1)=3$
$f(2)=9$	$g(2)=-1$
$f(5)=10$	$g(5)=-4$
$f(6)=13$	$g(6)=-6$

13. $(f+g)(5)$ 13. ______________

14. $(f-g)(2)$ 14. ______________

15. $(f \cdot g)(-1)$ 15. ______________

16. $\left(\frac{f}{g}\right)(6)$ 16. ______________

17. $(g \circ f)(-1)$ 17. ______________

Name: **Date:**
Instructor: **Section:**

Section 9.2 Inverse Functions

Learning Objectives
1. Determine whether a function is a one-to-one function.
2. Use the horizontal line test to decide whether a function is a one-to-one function.
3. Find the inverse of a function.
4. Find the equation of the inverse of a function.
5. Graph functions and their inverses.
6. Determine whether two functions are inverses of each other.

Vocabulary
Use the choices to complete each statement.

Horizontal	**Identical**	**Inverse of *f***
Reversed	**Vertical**	***x***
$y = x$	$\frac{1}{f}$	

1. The graphs of a function and its inverse are symmetric about the____________________ line.
2. The ordered pair $(5,4)$ in a function will be ____________________ in its inverse.
3. To tell if a function is one-to-one it must pass the ____________________ line test.
4. The symbol f^{-1} means ____________________.
5. Two functions are inverses if both $\left(f \circ f^{-1}\right)(x)$ and $\left(f^{-1} \circ f\right)(x)$ result to be ____________________.

Objective 1 and 3

Determine whether each function is one-to-one to function. If it is one-to-one, list the inverse function by switching the coordinates.

6. $f = \{(11,12),(4,3),(3,4),(6,6)\}$ 6. ____________________

7. $g = \{(3,2),(4,6),(5,-2),(-1,6)\}$ 7. ____________________

Name: **Date:**
Instructor: **Section:**

Objective 2

Determine whether the graph of each function is the graph of a one-to-one function.

8.

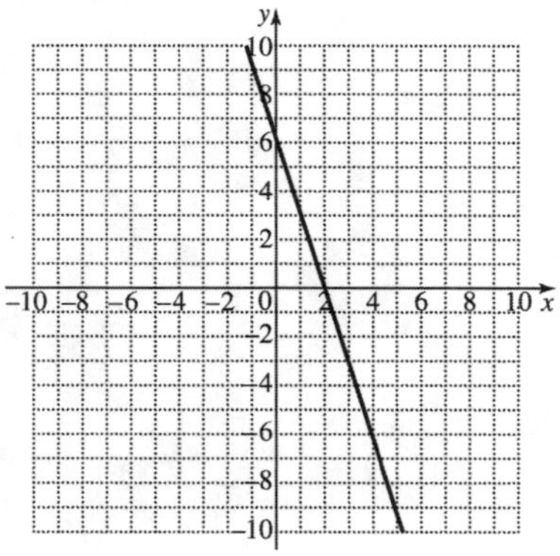

8. ______________________

9.

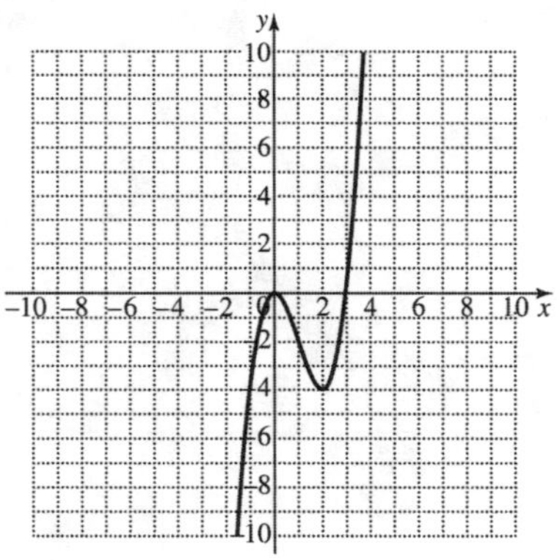

9. ______________________

Objective 4

Find the inverse of the following functions.

10. $f(x) = 2x - 3$

10. ______________________

Name: **Date:**
Instructor: **Section:**

11. $f(x)=\frac{x+3}{4}$ 11. ____________________

12. $f(x)=\frac{5}{3x+1}$ 12. ____________________

13. $f(x)=\sqrt[3]{x-5}$ 13. ____________________

Name: Date:
Instructor: Section:

Objective 5

Each of the following functions is one-to-one. Find the inverse of each function and graph the function and its inverse on the same set of axes.

14. $f(x) = 2x - 3$ 14. ____________________

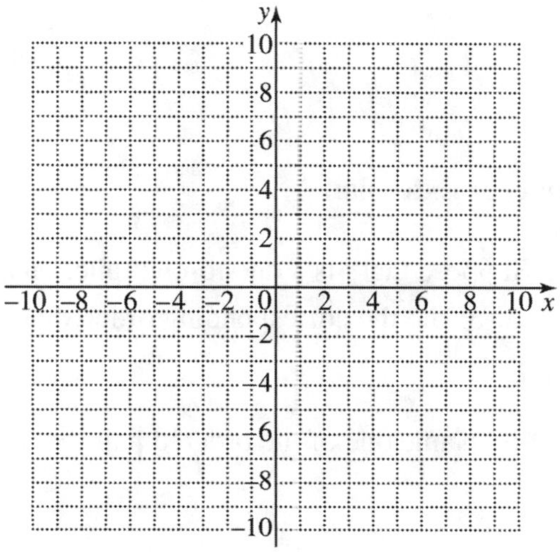

15. $f(x) = 5x + 1$ 15. ____________________

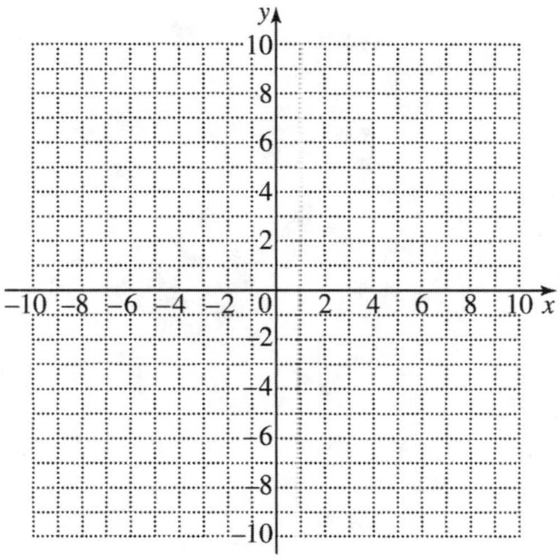

Name: **Date:**
Instructor: **Section:**

Objective 6

Solve.

16. If $f(x) = 3x + 2$, show that $f^{-1}(x) = \frac{x-2}{3}$.

16. ____________________

Concept Extension

17. Suppose that f is a one-to-one function and that $f(-5) = 25$.
    a. Write the corresponding ordered pair.
    b. Name one of the ordered pair that is a solution of f^{-1}

17a. ____________________

17b. ____________________

Section 9.3 Exponential Functions

Learning Objectives

1. Graph exponential functions.
2. Solve equations of the form $b^x = b^y$.
3. Solve the problems modeled by exponential equations.

Objective 1

Graph each exponential function.

1. $y = 2^x$

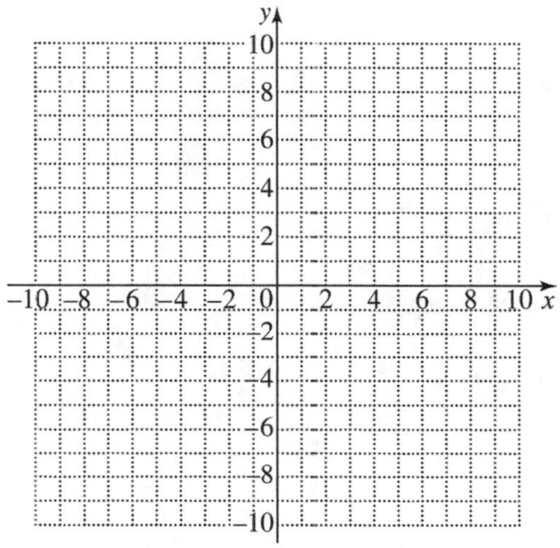

2. $y = \left(\frac{1}{4}\right)^x$

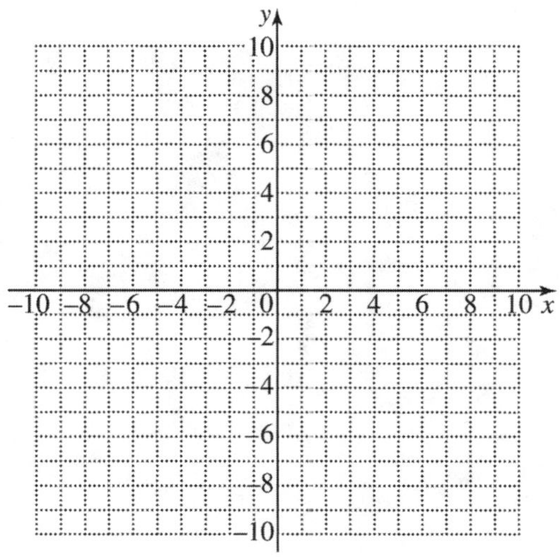

Name: **Date:**
Instructor: **Section:**

3. $y = 3^{x+2}$

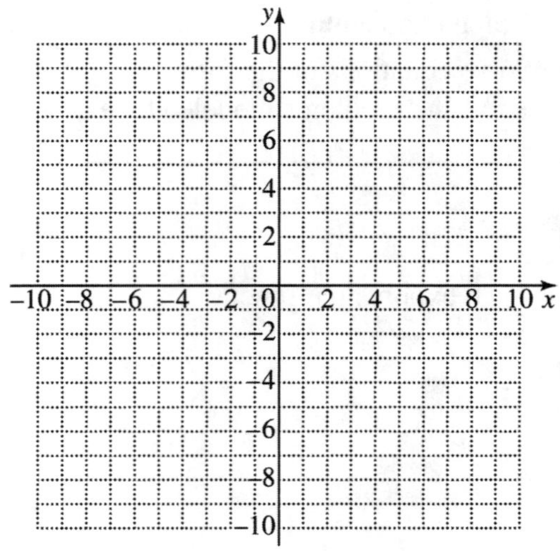

Objective 2

Solve each equation for x.

4. $2^x = 8$ 4. ____________________

5. $5^x = \frac{1}{125}$ 5. ____________________

6. $128 = 2^{6x}$ 6. ____________________

Name: **Date:**
Instructor: **Section:**

7. $27^{x+1} = 9$

7. ____________________

Solve. Unless otherwise indicated, round results to one decimal place.

8. One type of uranium has a daily radioactive decay rate of 0.4%. If 30 pounds of this uranium is available today, find how much will still remain after 50 days. Use $y = 30(2.7)^{-0.004t}$, and let t be 50.

8. ____________________

9. A bank account grows at 8% interest being compounded continuously. How much money will be in an account if it started with $25,000 and the money will sit in the account for 25 years. Use $A = 25000(2.7)^{.08t}$, and let t = 25.

9. ____________________

Name: **Date:**
Instructor: **Section:**

10. The decline of a population of rats is modeled by the equation $P = 8{,}000(2.7)^{-0.008t}$. How large will the population be if $t = 15$?

10. ______________________

Concept Extension

11. Graph the function: $y = 2^{|x|}$.

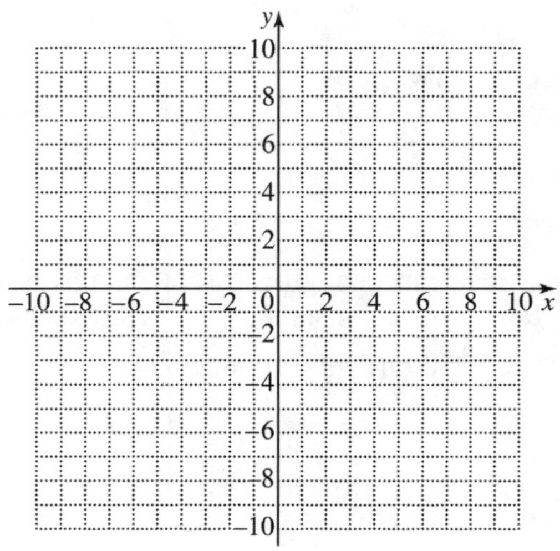

Name: **Date:**
Instructor: **Section:**

Section 9.4 Logarithmic Functions

Learning Objectives

1. Write exponential equations with logarithmic notation and write logarithmic equations with exponential notation.
2. Solve logarithmic equations by using exponential notation.
3. Identify and graph logarithmic functions.

Objective 1

Write each as an exponential equation.

1. $\log_4 16 = 2$ 1. ____________________

2. $\log_2 \frac{1}{8} = -3$ 2. ____________________

3. $\log_7 x = 9$ 3. ____________________

4. $\log_6 \sqrt[3]{6} = \frac{1}{3}$ 4. ____________________

Write each as a logarithmic function.

5. $5^{-3} = \frac{1}{125}$ 5. ____________________

6. $2^{1/4} = \sqrt[4]{2}$ 6. ____________________

Name: **Date:**
Instructor: **Section:**

7. $3^4 = 81$ 7. ______________

8. $\pi^7 = z$ 8. ______________

Find the value of each logarithmic expression.

9. $\log_2 32$ 9. ______________

10. $\log_{12} 1$ 10. ______________

11. $\log_6 \frac{1}{36}$ 11. ______________

Objective 2

Solve.

12. $\log_x 100 = 2$ 12. ______________

13. $\log_3 \frac{1}{27} = x$ 13. ______________

Name:
Instructor:

Date:
Section:

14. $\log_2 64 = x$

14. ______________________

15. $\log_5 5^8 = x$

15. ______________________

16. $6^{\log_6 9} = x$

16. ______________________

Simplify.

17. $2^{\log_2 3}$

17. ______________________

18. $\log_3 9$

18. ______________________

Objective 3

Graph each logarithmic function. Label any intercepts.

19. $y = \log_3 x$

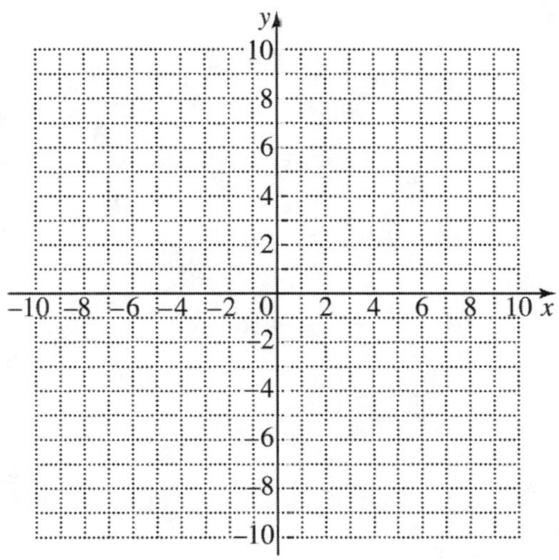

Name: **Date:**
Instructor: **Section:**

20. $f(x) = \log_5 x$

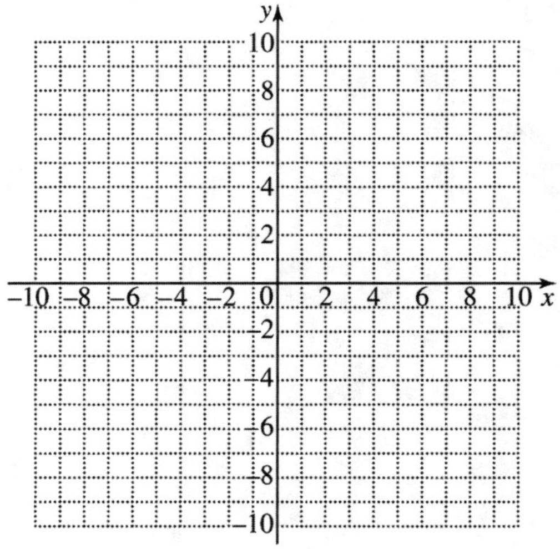

21. $f(x) = \log_{1/2} x$

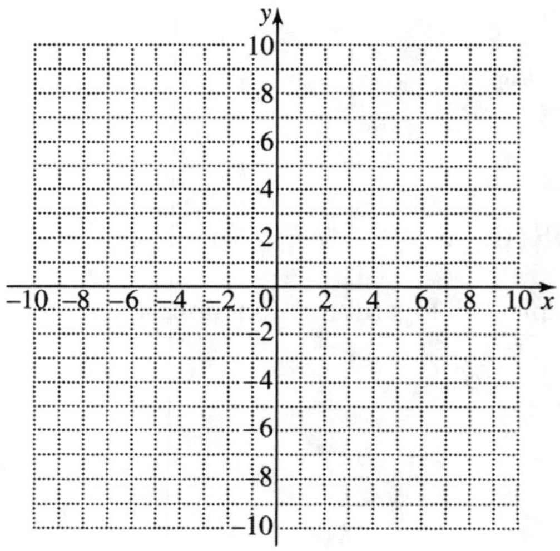

Name: **Date:**
Instructor: **Section:**

Concept Extension

Graph the following two functions on the same grid. Is there a relationship between the two of them?

22. $f(x) = \log_4 x$ $\qquad g(x) = 4^x$

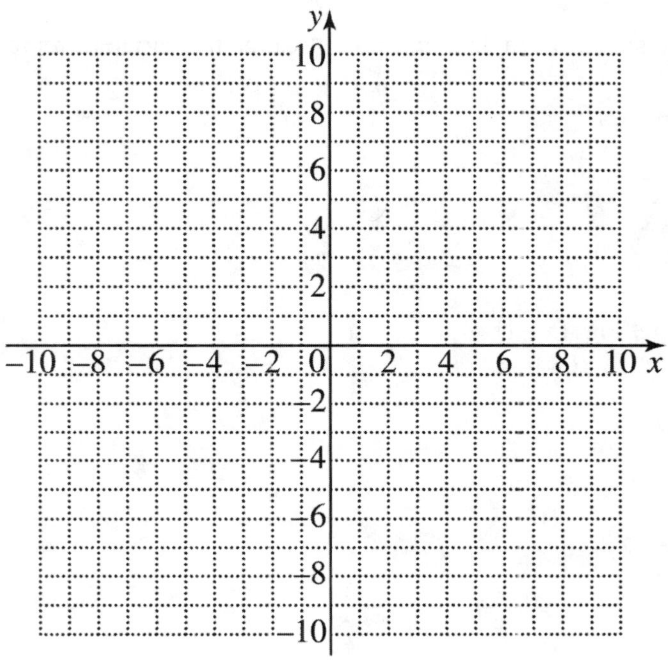

Name: **Date:**
Instructor: **Section:**

Section 9.5 Properties of Logarithms

Learning Objectives

1. Use the product property of logarithms.
2. Use the quotient property of logarithms.
3. Use the power property of logarithms.
4. Use the properties of logarithms together.

Objective 1

Write each sum as a single logarithm. Assume the variables represent positive numbers.

1. $\log_3 x + \log_3 5$ 1. ____________________

2. $\log_7 2x + \log_7 (x+1)$ 2. ____________________

3. $\log_7 (x-2) + \log_7 x^4$ 3. ____________________

4. $\log_{10} 5 + \log_{10} 2 + \log_{10} (x^2 + 2)$ 4. ____________________

Objective 2

Write each difference as a single logarithm. Assume that variables represent positive numbers.

5. $\log_3 x - \log_3 6$ 5. ____________________

Name: **Date:**
Instructor: **Section:**

6. $\log_{11} 5x - \log_{11}(x+6)$ 6. ____________________

7. $\log_2(x^2-6) - \log_2(x+5)$ 7. ____________________

8. $\log_9 150 - \log_9 25$ 8. ____________________

Objective 3

Use the power property to rewrite each expression.

9. $\log_4 x^6$ 9. ____________________

10. $\log_6 7^3$ 10. ____________________

11. $\log_5 \sqrt{y}$ 11. ____________________

12. $\log_8 10^{-5}$ 12. ____________________

Name: **Date:**
Instructor: **Section:**

Objective 4

Write each as a single logarithm.

13. $\log_4 2+\log_4 10-\log_4 5$ 13. ____________________

14. $\log_2(x+2)+3\log_2 x-\log_2(x-5)$ 14. ____________________

15. $4\log_8 x-3\log_8 x+2\log_8 2$ 15. ____________________

16. $\frac{1}{2}\log_6 x+4\log_6 x-2\log_6(x+2)$ 16. ____________________

Write each expression as a sum or difference of logarithms. Assume that variables represent positive numbers.

17. $\log_3 5x^2$ 17. ____________________

Name:
Instructor:

Date:
Section:

18. $\log_6 \frac{8y}{x^3}$

18. ____________________

19. $\log_7 \frac{\sqrt{x}}{(x+4)^2}$

19. ____________________

20. $\log_5 \sqrt[3]{\frac{4x}{7}}$

20. ____________________

Concept Extension

21. If $\log_b 3x = 1 + \log_b x$, find b.

21. ____________________

Name: **Date:**
Instructor: **Section:**

Section 9.6 Common Logarithms, Natural Logarithms, and Change of Base

Learning Objectives

1. Identify common logarithms and approximate them by calculator.
2. Evaluate common logarithms of powers of 10.
3. Identify natural logarithms and approximate them by calculator.
4. Evaluate natural logarithms of powers of *e*.
5. Use the change of base formula.

Objective 1

Use a calculator to approximate each logarithm to four decimal places.

1. $\log 2.56$ 1. ____________________

2. $\log 0.028$ 2. ____________________

Objective 2

Find the exact value.

3. $\log 1000$ 3. ____________________

4. $\log 0.0000001$ 4. ____________________

5. $\log \frac{1}{10,000}$ 5. ____________________

Solve each equation for x. Give an exact solution and a four-decimal-place approximation.

6. $\log x = 2.2$ 6. ____________________

7. $\log 2x = 1.1$ 7. ____________________

Name: **Date:**
Instructor: **Section:**

8. $\log(3x-2) = -0.3$ 8. ______________________

Objective 3

Use a calculator to approximate each logarithm to four decimal places.

9. $\ln 5$ 9. ______________________

10. $\ln 0.061$ 10. ______________________

Objective 4

Find the exact value.

11. $\ln e^{0.02}$ 11. ______________________

12. $\ln \sqrt[5]{e}$ 12. ______________________

13. $\ln \frac{1}{e^5}$ 13. ______________________

Name: Date:
Instructor: Section:

14. Find how much money Dana Jones has after 12 years if $1400 is invested at 8% interest compounded continuously. Use the formula $A = Pe^{rt}$ to solve.

14. ____________________

Objective 5

Approximate each logarithm to four decimal places.

15. $\log_4 9$

15. ____________________

16. $\log_{1/2} 12$

16. ____________________

17. $\log_6 0.023$

17. ____________________

Concept Extension

18. Without using a calculator, which of the following is larger? $\log 80$ or $\ln 80$

18. ____________________

Name: **Date:**
Instructor: **Section:**

Section 9.7 Exponential and Logarithmic Equations and Applications

Learning Objectives
1. Solve exponential equations.
2. Solve logarithmic equations.
3. Solve problems that can be modeled by exponential and logarithmic equations.

Objective 1

Solve each equation. Give an exact solution, and also approximate the solution to four decimal places.

1. $5^x = 9$ 1. ____________________

2. $2^{x-3} = 5$ 2. ____________________

3. $6^{3x} = 24$ 3. ____________________

4. $5^{2x-1} = 60$ 4. ____________________

Name:
Instructor:

Date:
Section:

Objective 2

Solve each equation.

5. $\log_3(x+5)=2$

5. ______________

6. $\log_4 x+\log_4(x+6)=2$

6. ______________

7. $\ln 8+\ln(x+1)=0$

7. ______________

8. $2\log x-\log x^3=4$

8. ______________

9. $\ln 4+\ln x=0$

9. ______________

Name: **Date:**
Instructor: **Section:**

Objective 3

Solve. Round answers to the nearest whole number.

10. The size of the wolf population as Isle Royale National Park increases at a rate of 4.3% per year. If the size of the current population is 83 wolves, find how many there should be in 5 years. Use $y = y_0 e^{0.043t}$ and round to the nearest whole.

10. ____________________

11. Find how long it will take $750 to triple if it is invested at 8% interest compounded quarterly. Use the formula $A = P\left(1+\frac{r}{n}\right)^{nt}$ to solve this compound interest problem.

11. ____________________

12. A bacterial culture grows according to the formula $P = P_0 a^t$. If it takes 4 days for the culture to double in size, how long will it take it to triple?

12. ____________________

Name: **Date:**
Instructor: **Section:**

Concept Extension

Without solving the following equations find the values of x that cannot be a solution.

13. $\log(x-5)-\log(x^2+7)=0$

13. ____________________

14. $\log(x^2+1)-\log(x+4)=0$

14. ____________________

Name:
Instructor:

Date:
Section:

Chapter 9 Vocabulary

Vocabulary Word	**Definition**	**Example**
The Algebra of Functions • Sum • Difference • Product • Quotient • Composite	• $(f+g)(x)=f(x)+g(x)$ • $(f-g)(x)=f(x)-g(x)$ • $(f\cdot g)(x)=f(x)\cdot g(x)$ • $\left(\frac{f}{g}\right)(x)=\frac{f(x)}{g(x)}$ • $(f\circ g)(x)=f(g(x))$	$f(x)=3x \qquad g(x)=x+2$ • $(f+g)(x)=3x+x+2=4x+2$ • $(f-g)(x)=3x-(x+2)=2x-2$ • $(f\cdot g)(x)=3x\cdot(x+2)=3x^2+6x$ • $\left(\frac{f}{g}\right)(x)=\frac{3x}{x+2}$ • $(f\circ g)(x)=3(x+2)=3x+6$
One-to-one function	A function that not only for ever x there is only one y, but for every y there is only one x. Must pass BOTH vertical and horizontal line test.	$y=3x+2$ is a one-to-one function.
Inverse function	The inverse of f is denoted as f^{-1}. Ordered pairs are reversed in the inverse function.	$y=4x+3$ and $y=\frac{x-3}{4}$ are inverses.
Exponential function	$f(x)=b^x$; $b>0,\ b\neq 0$	$f(x)=7^x$
Logarithmic function	$f(x)=\log_b x$; $b>0,\ b\neq 1$	$f(x)=\log_3 5$
Properties of logarithms • Product Property • Quotient Property • Power Property	• $\log_b xy=\log_b x+\log_b y$ • $\log_b \frac{x}{y}=\log_b x-\log_b y$ • $\log_b x^r=r\log_b x$	• $\log_2 5y=\log_2 5+\log_2 y$ • $\log_2 \frac{6x}{7}=\log_2 6x-\log_2 7$ • $\log_4 x^6=6\log_4 x$
Common Logarithm	$\log x=\log_{10} x$	

Name: **Date:**
Instructor: **Section:**

Natural Logarithm	$\ln x = \log_e x$	
Continuously Compound Interest Formula	$A = Pe^{rt}$; r is the annual interest rate for P dollars invested for t years	
Logarithm Property of Equality	$\log_b a = \log_b c$ then $a = c$	$\log_4 x = \log_4 5$ $x = 5$

Name: **Date:**
Instructor: **Section:**

Practice Test A

If $f(x)=3x-4$ and $g(x)=-2x$, find the following.

1. $(f+g)(x)$ 1. ____________

2. $(f \cdot g)(x)$ 2. ____________

3. $(f \circ g)(x)$ 3. ____________

4. $(g \circ f)(4)$ 4. ____________

Find the inverse of the following functions.

5. $f(x)=\frac{1}{2}x-7$ 5. ____________

Name:
Instructor:

Date:
Section:

6. $f(x) = x^3 - 4$

6. ______________________

7. $f(x) = \frac{5x+8}{7}$

7. ______________________

Use the properties of logarithms to write each expression as a single logarithm.

8. $2\log_6 x + \log_6 3 - \log_6 (x+2)$

8. ______________________

9. $\frac{1}{2}\log_3 x - 3\log_3 2 + \log_3 5$

9. ______________________

10. Approximate $\log_5 12$ to the four decimal places.

10. ______________________

Name: **Date:**
Instructor: **Section:**

Solve each logarithmic equation for x.

11. $\log_4 x = 5$ 11. ______________

12. $\ln \sqrt[5]{e} = x$ 12. ______________

13. $\log_2 \left(x^2 - 2x\right) = 3$ 13. ______________

14. $\log_3 (x+5) - \log_3 (x-2) = 2$ 14. ______________

15. Solve $\ln(2x-5) = 1.68$ accurate to four decimal places. 15. ______________

Name:
Instructor:

Date:
Section:

Graph.

16. $y=\left(\frac{1}{5}\right)^x$

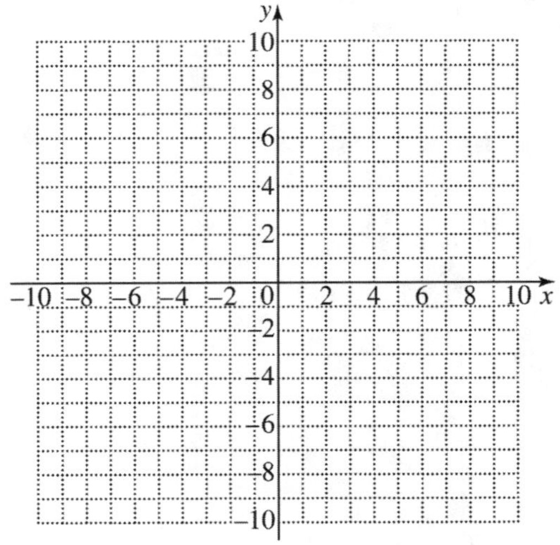

17. $y=4^x$

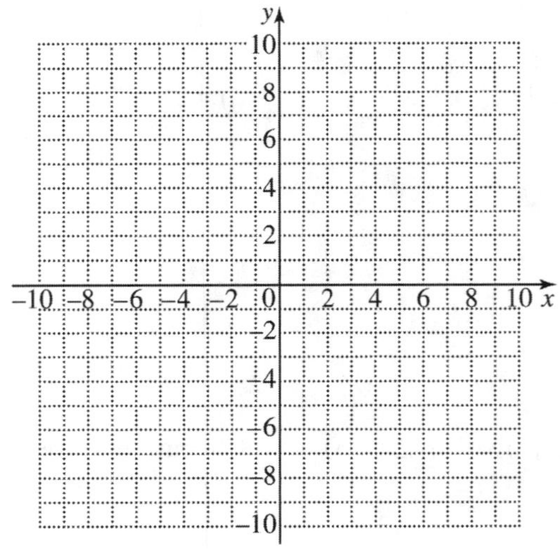

Name:
Instructor:

Date:
Section:

Solve.

18. Find the amount in the account if \$6000 is invested at 8.5% interested compounded quarterly for 7 years. Use the formula $A = P\left(1+\frac{r}{n}\right)^{nt}$.

18. ____________________

19. Find the amount of money that needs to be put in an account at 6% compounded continuously, if it is necessary to have \$12,000 in 3 years. Use the formula $A = Pe^{rt}$

19. ____________________

20. A radioactive material decays according to the formula $A = A_0(2)^{-t}$. How much of a 6-gram sample will be left in 4 years.

20. ____________________

Name: **Date:**
Instructor: **Section:**

Practice Test B

If $f(x) = 6x$ and $g(x) = 4 - 2x$, find the following.

1. $(f - g)(x)$ 1. ______________

a. $4x + 4$ b. $8x - 4$

c. $4 - 8x$ d. $4x - 4$

2. $(f \cdot g)(x)$ 2. ______________

a. $24x - 2$ b. $4 - 12x^2$

c. $24x - 12x^2$ d. $24x - 2x^2$

3. $(f \circ g)(x)$ 3. ______________

a. $24 - 12x$ b. $4 - 12x$

c. $4 - 12x^2$ d. $24x - 12x^2$

4. $(g \circ f)(-2)$ 4. ______________

a. 28 b. 32

c. 42 d. 48

Name:
Instructor:

Date:
Section:

Find the inverse of the following functions.

5. $f(x) = 4x + 9$

5. ____________________

a. $\dfrac{x}{4} - 9$

b. $4x - 9$

c. $\dfrac{x}{9} - 4$

d. $\dfrac{x - 9}{4}$

6. $f(x) = \sqrt[3]{x + 2}$

6. ____________________

a. $\sqrt[3]{x} + 2$

b. $\sqrt[3]{x - 2}$

c. $x^3 - 2$

d. $\sqrt[3]{x} - 2$

7. $f(x) = \dfrac{5 - 3x}{2}$

7. ____________________

a. $\dfrac{2x + 3}{-5}$

b. $\dfrac{2x - 5}{-3}$

c. $\dfrac{2x - 5}{3}$

d. $\dfrac{2x + 5}{3}$

Name: **Date:**
Instructor: **Section:**

Use the properties of logarithms to write each expression as a single logarithm.

8. $\log_3 6 - 2\log_3 3 + 5\log_3 x$ 8. ____________________

a. $\log_3 \frac{10x}{3}$ b. $\log_3 \frac{2}{3x^5}$

c. $\log_3 \frac{2x^5}{3}$ d. $\log_3 \frac{3x^5}{2}$

9. $\log_5 (x-1) - 6\log_5 x + \frac{1}{2}\log_5 2$ 9. ____________________

a. $\log_5 \frac{\sqrt{2}(x-1)}{x^6}$ b. $\log_5 \frac{(x-1)}{6x}$

c. $\log_5 \frac{6x}{\sqrt{2}(x-1)}$ d. $\log_5 \frac{x^6}{\sqrt{2}(x-1)}$

10. Approximate $\log_8 0.025$ to the four decimal places. 10. ____________________

a. – 1.2763 b. – 1.7740

c. 1.6742 d. 2.3321

Name:
Instructor:

Date:
Section:

Solve each logarithmic equation for x.

11. $\log_x 64 = 6$

11. ____________________

a. -4 b. 4

c. 2 d. -2

12. $\log x = 1.5$

12. ____________________

a. 31.6228 b. -24.5672

c. 32.3498 d. 28.7429

13. $\log_6(3x+8) = 2$

13. ____________________

a. 11 b. $-\frac{8}{3}$

c. -12 d. $\frac{28}{3}$

14. $\log_2 x - \log_2(x-2) = 4$

14. ____________________

a. $\frac{8}{3}$ b. $\frac{32}{15}$

c. $\frac{11}{5}$ d. 2

Name: **Date:**
Instructor: **Section:**

15. Solve $\ln(4x+5)=0.025$ accurate to four decimal places. 15. ______________

a. -0.9937 b. -0.9832

c. 0.9928 d. 0.9952

Graph.

16. $y=3^x$ 16. ______________

a.

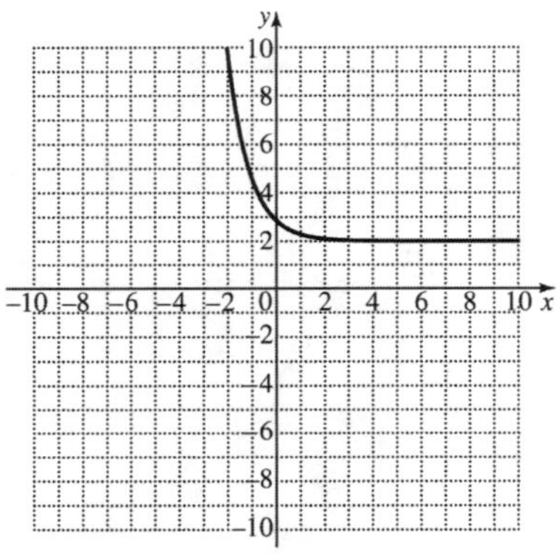

b.

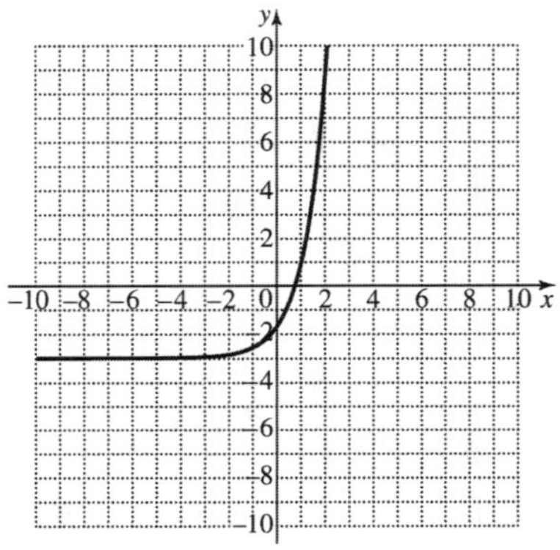

c.

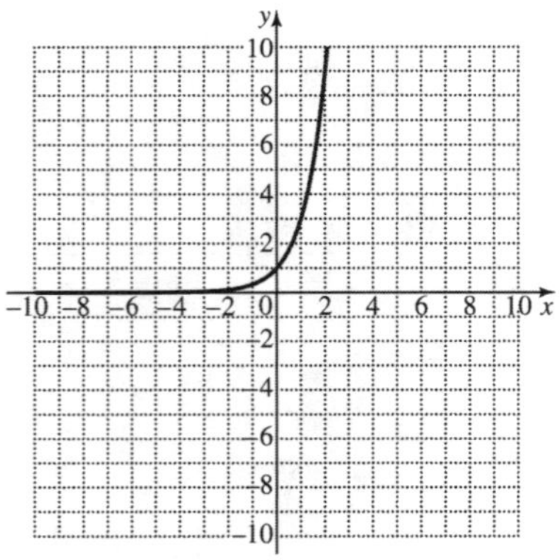

d.

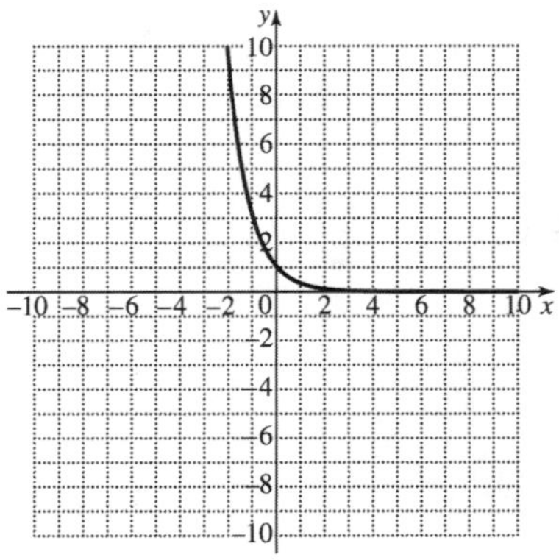

Name:
Instructor:

Date:
Section:

17. $y = \left(\frac{1}{2}\right)^x$

17. ______________________

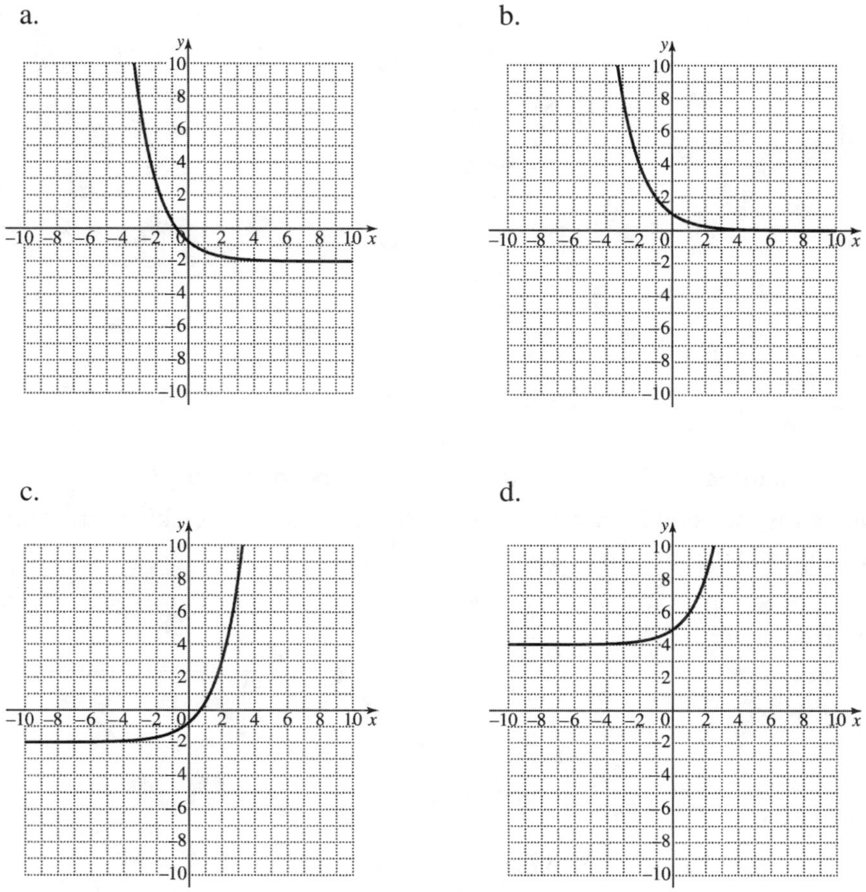

Solve.

18. Find the amount in the account if $4500 is invested at 6% interested compounded quarterly for 10 years. Use the formula $A = P\left(1+\frac{r}{n}\right)^{nt}$.

18. ______________________

a. $8214.46

b. $8220.34

c. $8127.50

d. $8163.08

19. Find the amount of money to which a $1250 investment will grow to if it is invested at 9.5% compounded continuously for 5 years. Use the formula $A = Pe^{rt}$

19. ____________________

a. $1899.43   b. $1998.67

c. $2010.02   d. $2105.34

20. The spread of foot and mouth disease through a herd of cattle can be modeled using the formula $P = P_0e^{0.25t}$, where t is the number of days. If a rancher does not act quickly to treat three cases, how many cattle will have the disease in 2 weeks?

20. ____________________

a. 99 cows   b. 95 cows

c. 102 cows   d. 89 cows

Name:
Instructor:

Date:
Section:

Chapter 10 Conic Sections
Section 10.1 The Parabola and the Circle

Learning Objectives

1. Graph parabolas of the form $x = a(y-k)^2 + h$ and $y = a(x-h)^2 + k$.
2. Graph circles of the form $(x-h)^2 + (y-k)^2 = r^2$.
3. Write the equation of a circle, given its center and radius.
4. Find the center and the radius of a circle, given its equation.

Vocabulary

Use the choices to complete each statement.

Center	**Circle**	**Conic Sections**
Diameter	**Radius**	**Vertex**

1. Half a circle's diameter is its ______________________.
2. For a parabola that opens downward, the highest point is called the ______________________.
3. The circle and parabola are examples of ______________________.
4. The distance from one side of the circle to the other, running through the center, is the

   __________________.

5. A ____________________ is the set of all points equidistant for a fixed point called the

   __________________ .

Objective 1

The graph of each equation is a parabola. Find the vertex of the parabola and sketch the graph.

6. $x = (y-2)^2 + 3$

6. ______________________

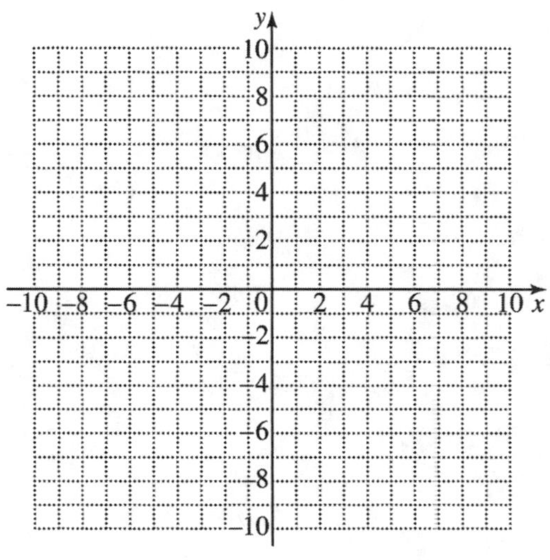

Name:
Instructor:

Date:
Section:

7. $y = -2(x+1)^2 - 4$

7. ______________________

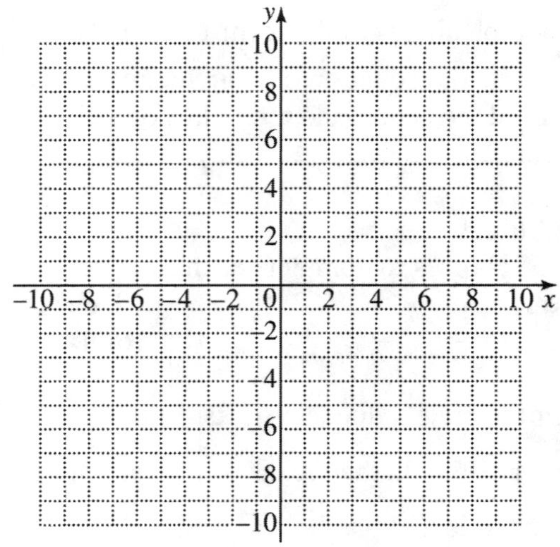

8. $x = 3y^2 + 6y + 2$

8. ______________________

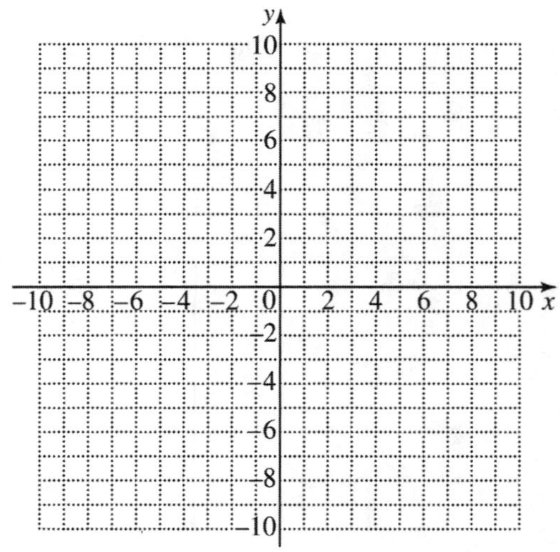

9. $y = x^2 - 6x + 3$

9. ____________________

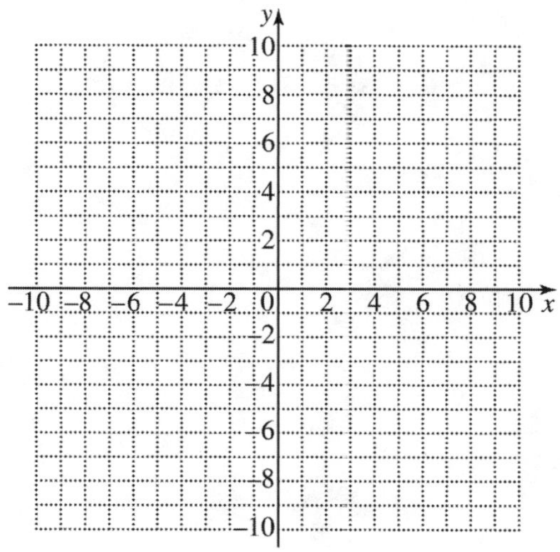

Objective 2

The graph of each equation is a circle. Find the center and the radius, and then sketch.

10. $(x-5)^2 + (y+2)^2 = 1$

10. ____________________

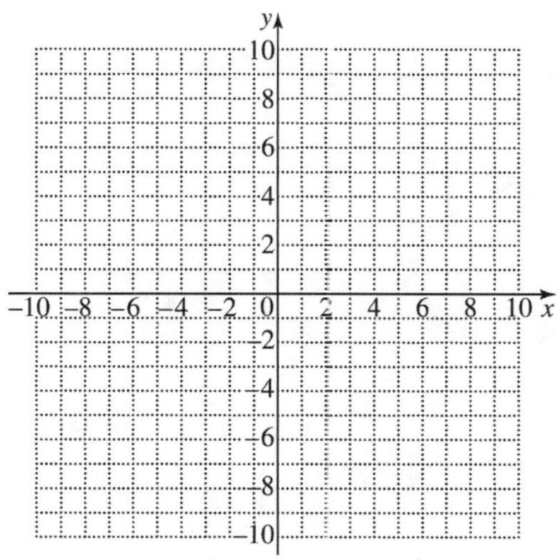

Name: **Date:**
Instructor: **Section:**

11. $x^2 + (y-3)^2 = 4$

11. ____________________

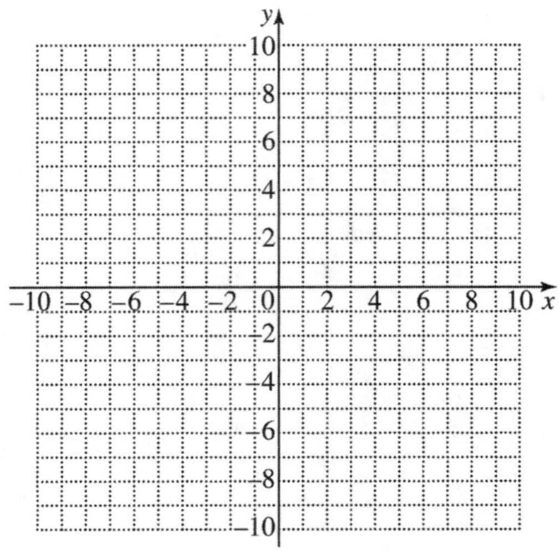

12. $x^2 + 6x + y^2 - 8y = 0$

12. ____________________

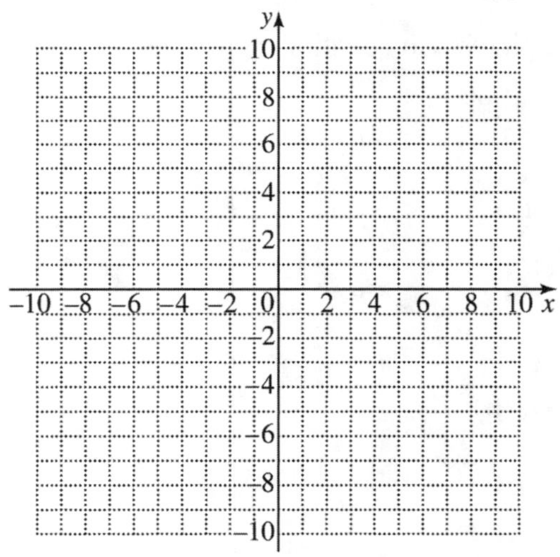

Name:
Instructor:

Date:
Section:

13. $x^2 + y^2 + 4x + 2y = 11$

13. ____________________

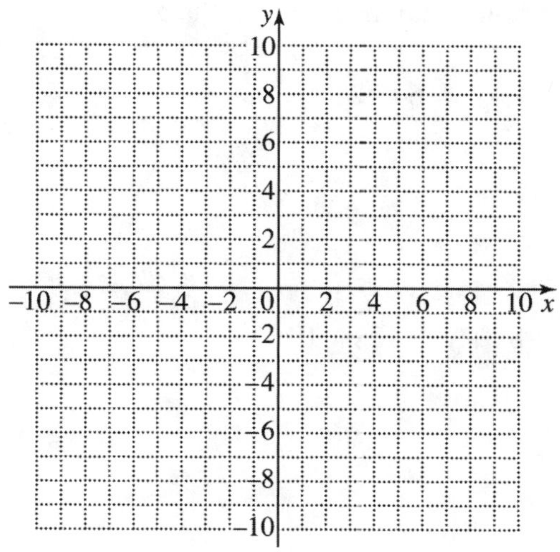

Objective 3

Write an equation of the circle with the given center and radius.

14. $(3,4)$; 5

14. ____________________

15. $(-1,-6)$; 7

15. ____________________

16. $(-5,4)$; $3\sqrt{5}$

16. ____________________

Name: **Date:**
Instructor: **Section:**

Objective 4

Find the center and the radius of the given circle equations.

17. $4x^2 + 4y^2 = 64$

17. ____________________

18. $x^2 + 8x + y^2 + 12y = 10$

18. ____________________

19. $x^2 + y^2 - 6x - 10y = 4$

19. ____________________

Concept Extension

20. Analyze the problems from this section, how can you tell by first glance if the equation is a parabola or a circle.

20. ____________________

21. Under what conditions would the graph of the $x = a(y-k)^2 + h$ have no y-intercepts.

21. ____________________

Name: **Date:**
Instructor: **Section:**

Name:
Instructor:

Date:
Section:

Section 10.2 The Ellipse and the Hyperbola

Learning Objective
1. Define and graph an ellipse.
2. Define and graph a hyperbola.

Vocabulary
Use the choices to complete each statement.

Center	**Ellipse**	**Foci**
Hyperbola	$(0,a)$ **and** $(0,-a)$	$(b,0)$ **and** $(-b,0)$

1. Two fixed points of a hyperbola or ellipse are referred to as ________________.
2. A(n) ________________ is a set of points such that the sum of their distances to the foci is a constant.
3. The graph $\frac{x^2}{a^2}-\frac{y^2}{b^2}=1$ is a(n) ________________ with the center at the origin and x-intercepts at ________________.
4. The ________________ is the midpoint between the two foci.

Objective 1

Sketch the graph of each equation.

5. $x^2+\frac{y^2}{4}=1$

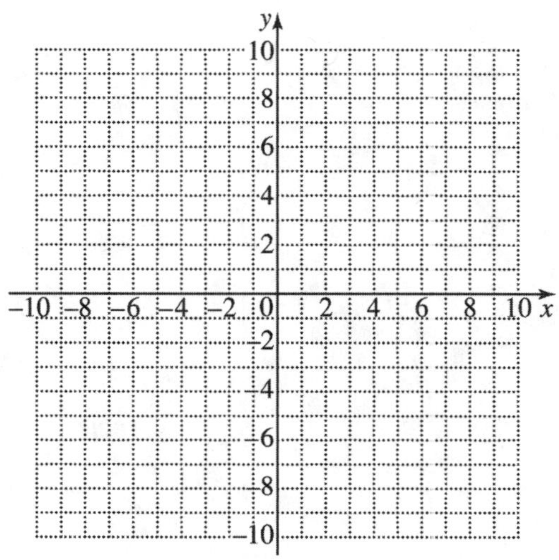

Name:
Instructor:

Date:
Section:

6. $\dfrac{x^2}{16}+\dfrac{y^2}{25}=1$

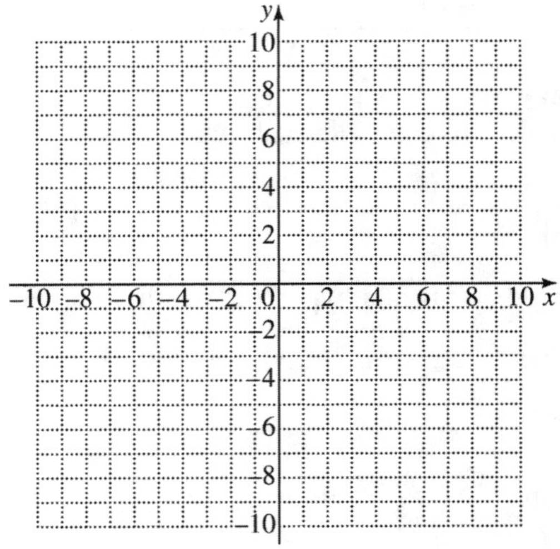

7. $16x^2 + y^2 = 64$

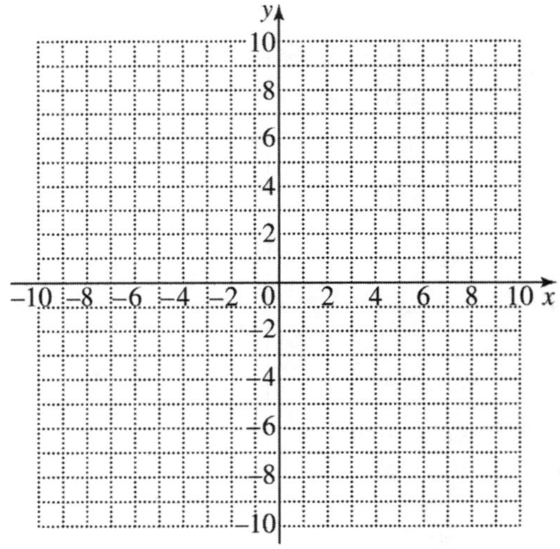

Name:
Instructor:

Date:
Section:

8. $\dfrac{(x-1)^2}{4}+\dfrac{(y-1)^2}{25}=1$

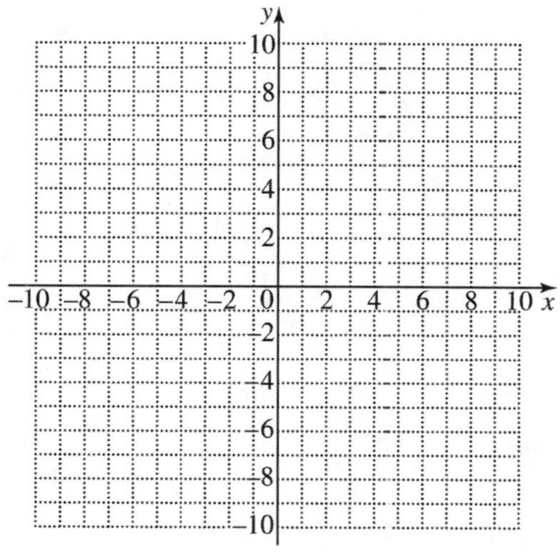

9. $\dfrac{(x+4)^2}{36}+\dfrac{(y+2)^2}{49}=1$

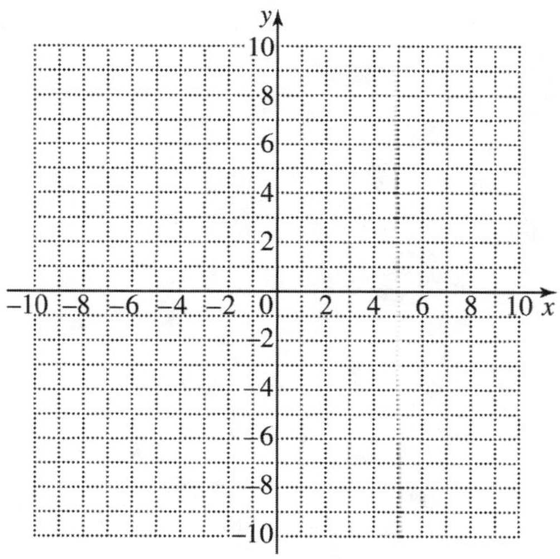

Name:
Instructor:

Date:
Section:

Objective 2

Sketch the graph of each equation.

10. $\dfrac{x^2}{4} - y^2 = 1$

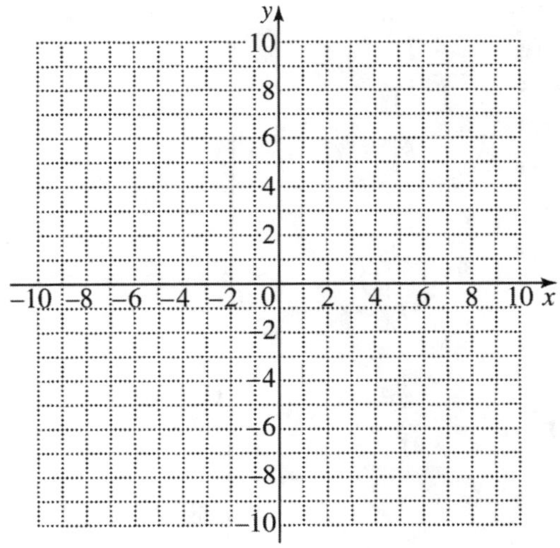

11. $\dfrac{x^2}{25} - \dfrac{y^2}{16} = 1$

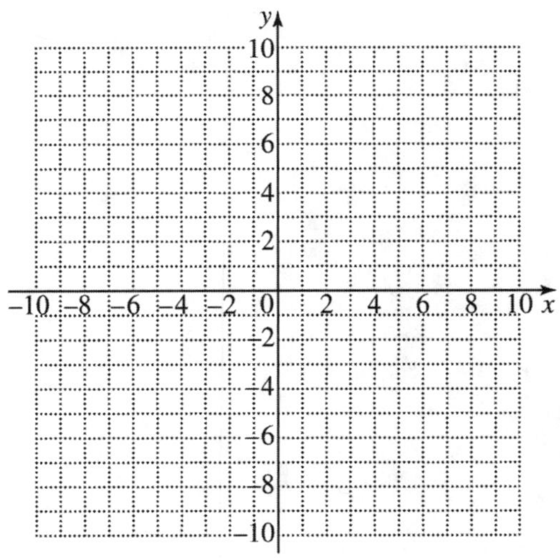

Name:
Instructor:

Date:
Section:

12. $9y^2 - x^2 = 36$

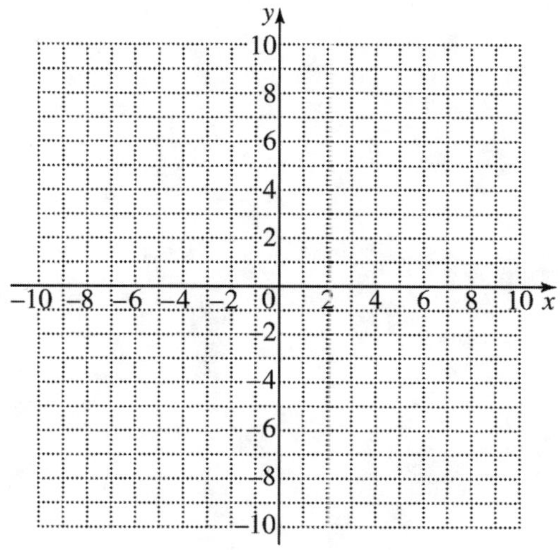

13. $\frac{(x-2)^2}{25} - \frac{(y-1)^2}{16} = 1$

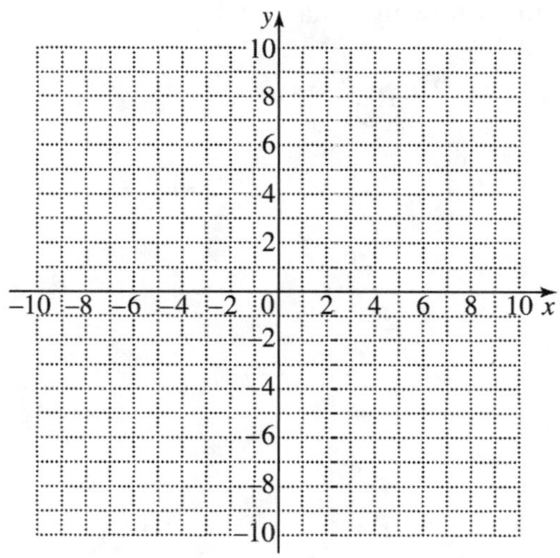

Name: **Date:**
Instructor: **Section:**

14. $\frac{(y+1)^2}{4} - \frac{(x+2)^2}{9} = 1$

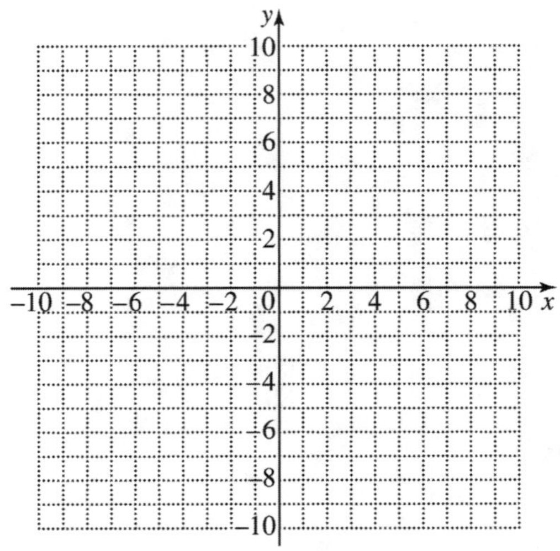

Concept Extension

Graph the following equations.

15. $4x^2 + 18y^2 - 16x - 36y = 2$

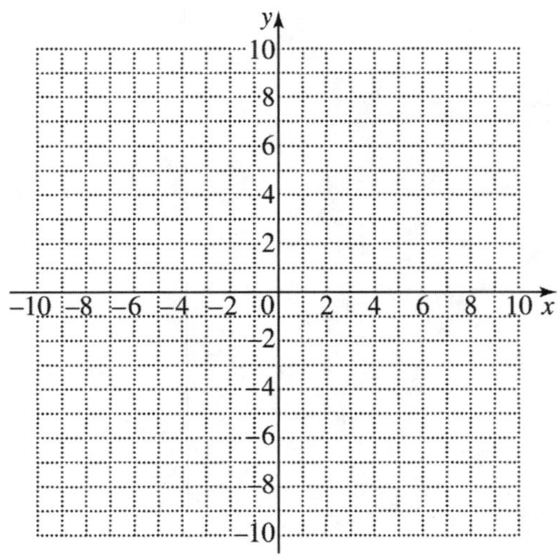

Name:
Instructor:

Date:
Section:

16. $x^2 + 4y - y^2 - 2x = 15$

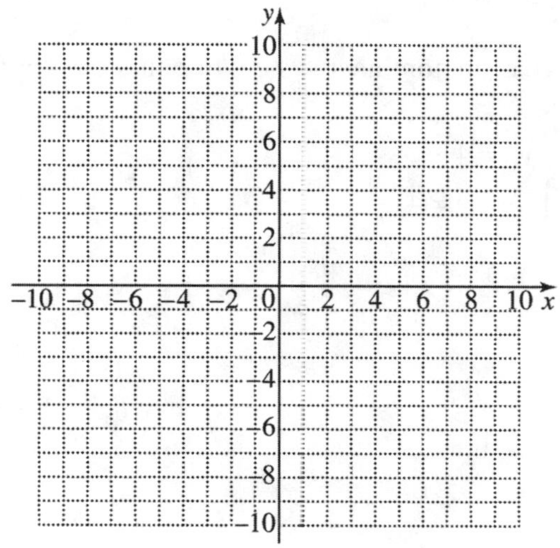

Name: **Date:**
Instructor: **Section:**

Section 10.3 Solving Nonlinear Systems of Equations

Learning Objectives
1. Solve a nonlinear system by substitution.
2. Solve a nonlinear system by elimination.

Objective 1

Solve each nonlinear system of equations for real solutions using substitution.

1. $\begin{cases} y^2 = 4 - x \\ x - 2y = 4 \end{cases}$

1. ______________________

2. $\begin{cases} x^2 + y^2 = 16 \\ 2x + 3y = 0 \end{cases}$

2. ______________________

3. $\begin{cases} 3x + y = 4 \\ xy = 2 \end{cases}$

3. ______________________

Name:
Instructor:

Date:
Section:

4. $\begin{cases} y = x^2 + 25 \\ y = 2x^2 + 9 \end{cases}$

4. ____________________

5. $\begin{cases} x^2 + 4y^2 = 16 \\ x - y = 4 \end{cases}$

5. ____________________

Objective 2

Solve each nonlinear system of equations for real solutions using elimination.

6. $\begin{cases} 2x^2 + 3y^2 = 14 \\ -x^2 + y^2 = 3 \end{cases}$

6. ____________________

Name: **Date:**
Instructor: **Section:**

7. $\begin{cases} y = x^2 + 12 \\ y = -x^2 + 10 \end{cases}$

7. ______________________

8. $\begin{cases} x^2 + y^2 = 36 \\ y = x^2 - 16 \end{cases}$

8. ______________________

9. $\begin{cases} x^2 + 4y^2 = 16 \\ x^2 + (y-1)^2 = 9 \end{cases}$

9. ______________________

Name:
Instructor:

Date:
Section:

10. $\begin{cases} x + y = 4 \\ y = x^2 + 4 \end{cases}$

10. ____________________

Concept Extension

Solve.

11. The sum of the squares of two numbers is 36. The sum of the first number and twice the second number is 6. Find the two numbers.

11. ____________________

12. How many possible real solutions are there for the intersection of a circle and an ellipse?

12. ____________________

Name: **Date:**
Instructor: **Section:**

Section 10.4 Nonlinear Inequalities and Systems of Inequalities

Learning Objectives
1. Graph a nonlinear inequality.
2. Graph a system of nonlinear inequalities.

Objective 1

Graph each inequality.

1. $\frac{x^2}{4}+\frac{y^2}{9}\le 1$

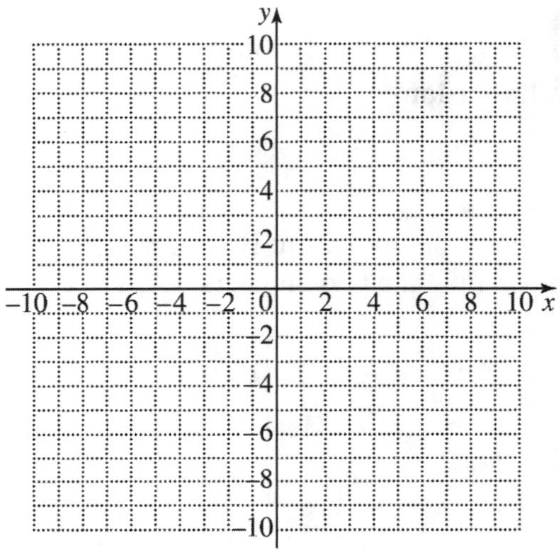

2. $\frac{x^2}{16}-y^2>1$

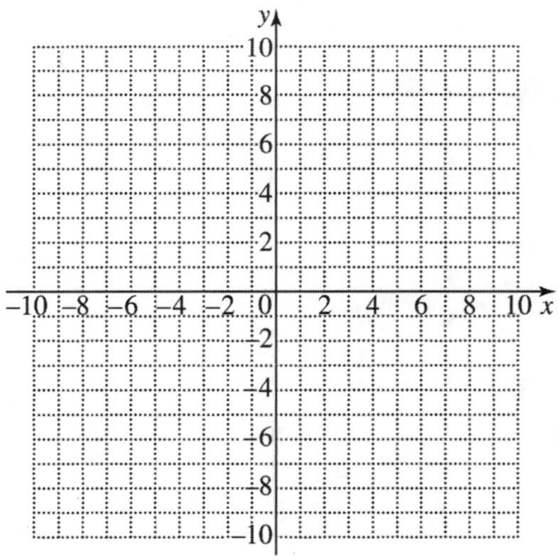

Name:
Instructor:

Date:
Section:

3. $x^2 + (y-2)^2 < 36$

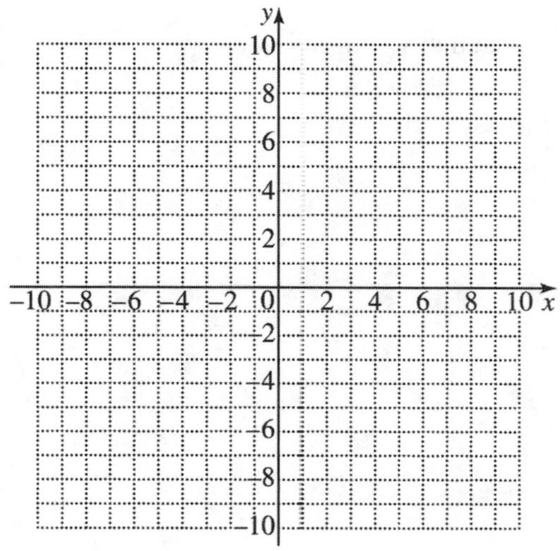

4. $y > x^2 + 2x$

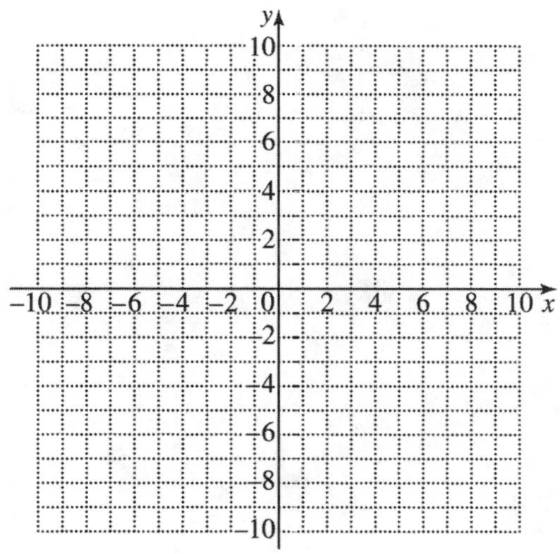

Name: **Date:**
Instructor: **Section:**

Objective 2

Graph each system.

5. $\begin{cases} x^2 + y^2 > 9 \\ y > x^2 \end{cases}$

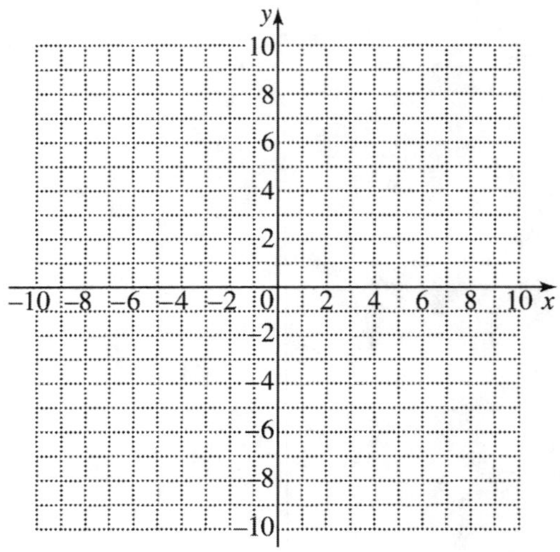

6. $\begin{cases} x^2 - y^2 \geq 1 \\ x + y < 2 \end{cases}$

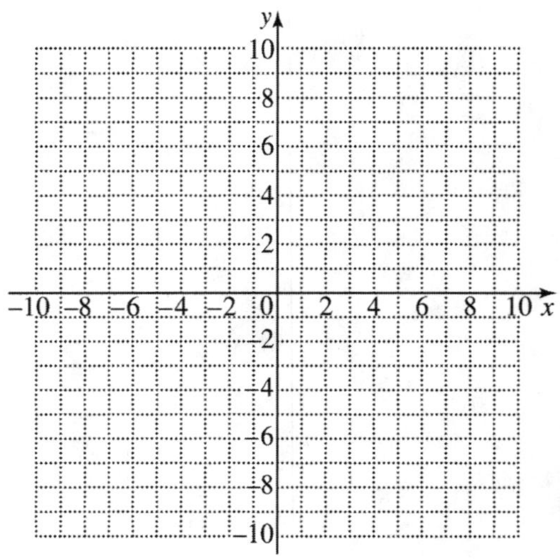

Name:
Instructor:

Date:
Section:

7. $\begin{cases} \dfrac{x^2}{16} + \dfrac{y^2}{25} > 1 \\ y < x^2 \end{cases}$

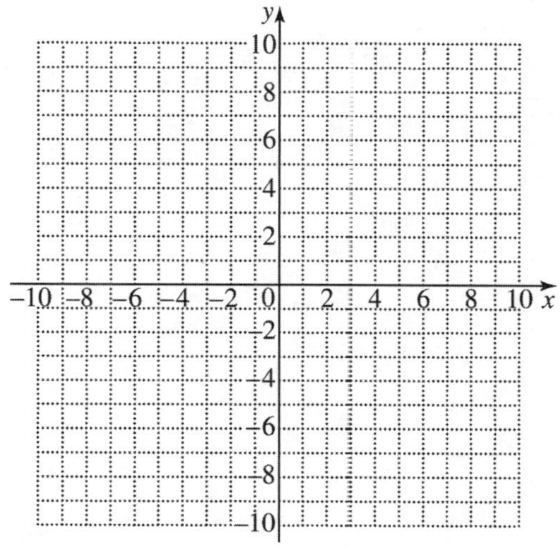

8. $\begin{cases} x^2 + y^2 \le 25 \\ x^2 + y^2 > 4 \end{cases}$

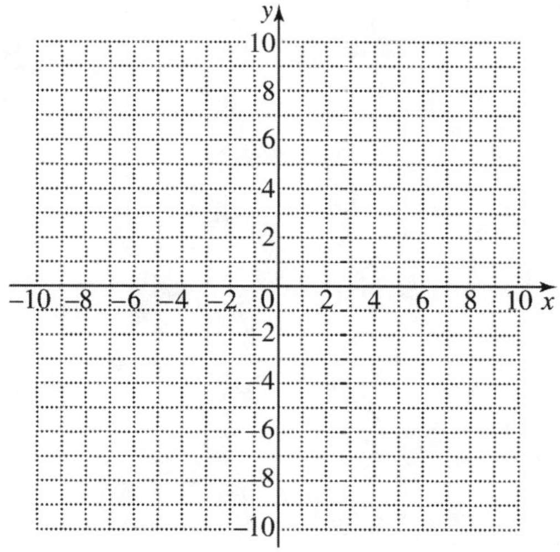

Name: **Date:**
Instructor: **Section:**

Concept Extension

Graph the following system.

9.
$$\begin{cases} y \geq x^2 \\ y < 2x + 1 \\ x \leq 3 \\ y < 4 \end{cases}$$

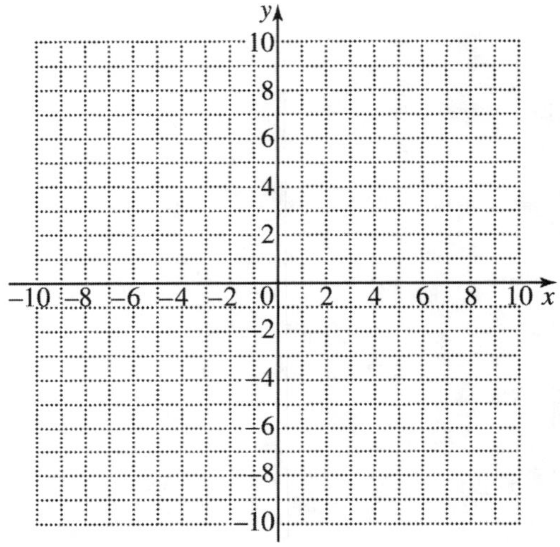

Name: Date:
Instructor: Section:

Chapter 10 Vocabulary

Vocabulary Word	Definition	Example
Parabolas	$y = a(x-h)^2 + k$ $x = a(y-k)^2 + h$	
Circle	$(x-h)^2 + (y-k)^2 = r^2$	
Ellipse	$\frac{x^2}{a^2} + \frac{y^2}{b^2} = 1$	
Hyperbola	$\frac{x^2}{a^2} - \frac{y^2}{b^2} = 1$ or $\frac{x^2}{b^2} - \frac{y^2}{a^2} = 1$	
Nonlinear System of Equations	A system of at least one that is not linear.	

Name:
Instructor:

Date:
Section:

Practice Test A

Sketch the graph of each equation.

1. $x^2 + y^2 = 64$

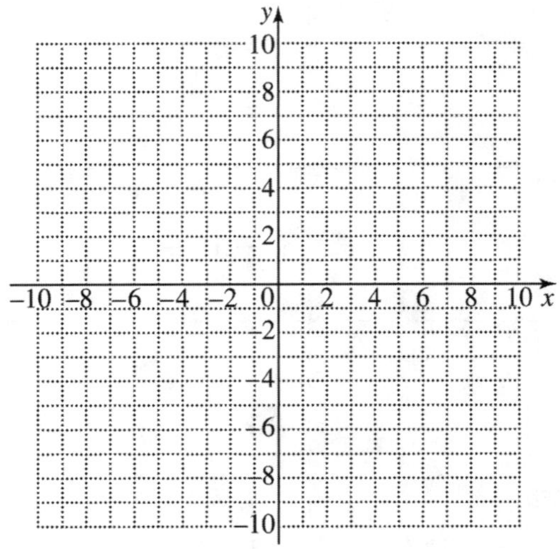

2. $x^2 - \frac{y^2}{16} = 1$

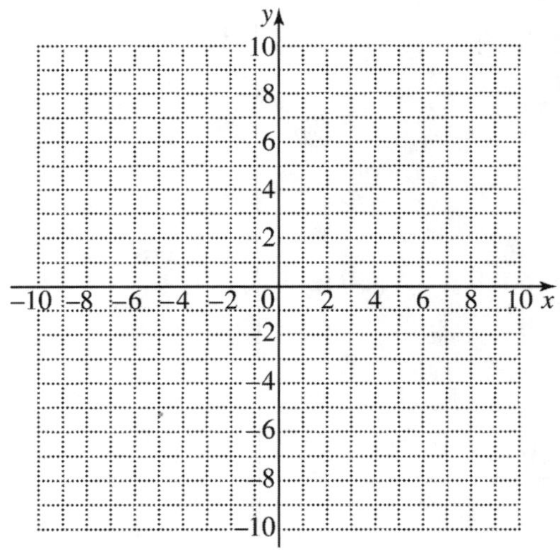

Name:
Instructor:

Date:
Section:

3. $y=(x-4)^2+6$

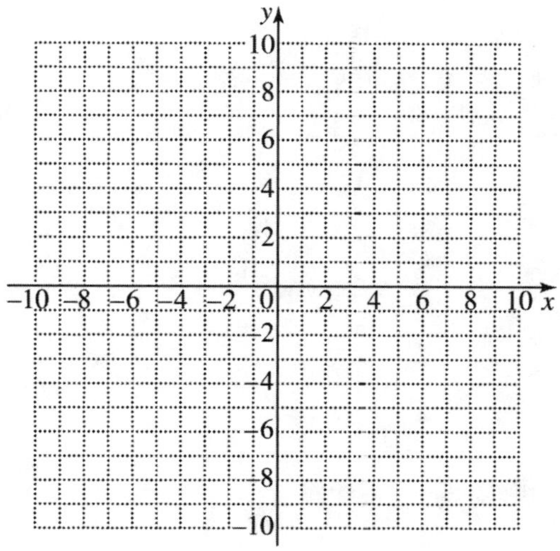

4. $\frac{(x+1)^2}{25}+\frac{(y-2)^2}{4}=1$

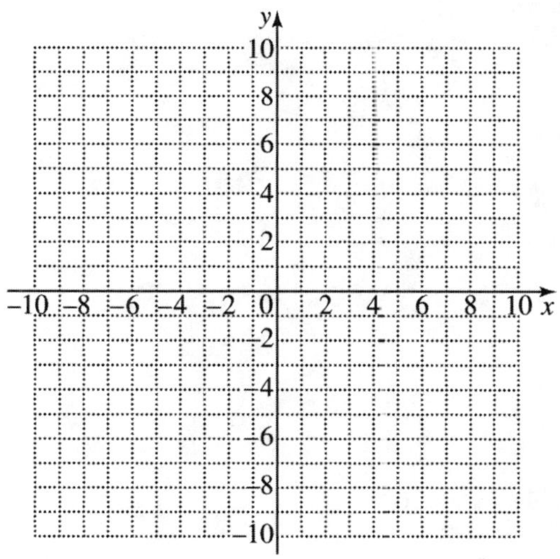

Name: **Date:**
Instructor: **Section:**

5. $x^2 + 8x - y^2 - 4y = 4$

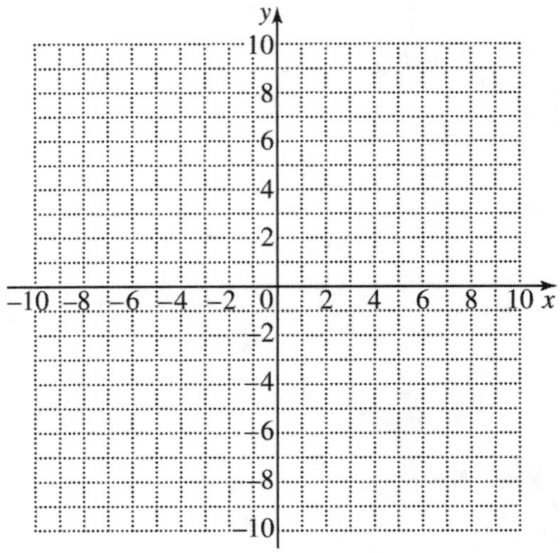

Solve each system.

6. $\begin{cases} x^2 - y^2 = 16 \\ 2x^2 + y^2 = 5 \end{cases}$

6. ____________________

7. $\begin{cases} 4x^2 + 4y^2 = 16 \\ x = y + 2 \end{cases}$

7. ____________________

Name: **Date:**
Instructor: **Section:**

8. $\begin{cases} y^2 - x^2 = 4 \\ x = 3y \end{cases}$

8. ______________________

Graph the solution of each system.

9. $\begin{cases} 3x + 4y > 12 \\ y < x^2 + 1 \end{cases}$

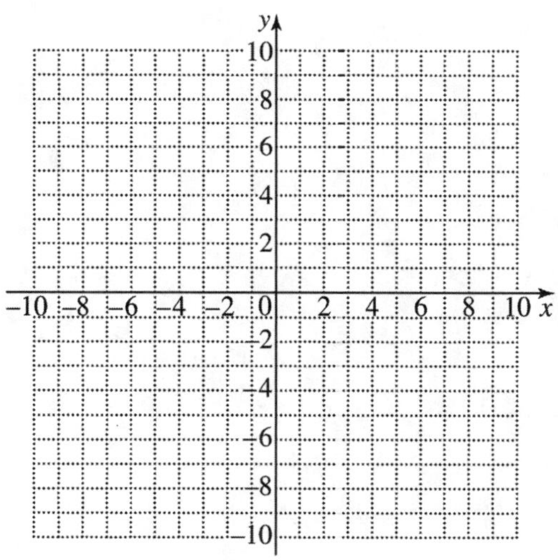

Name: **Date:**
Instructor: **Section:**

10. $\begin{cases} \dfrac{x^2}{16} - y^2 < 1 \\ x + y \geq 3 \end{cases}$

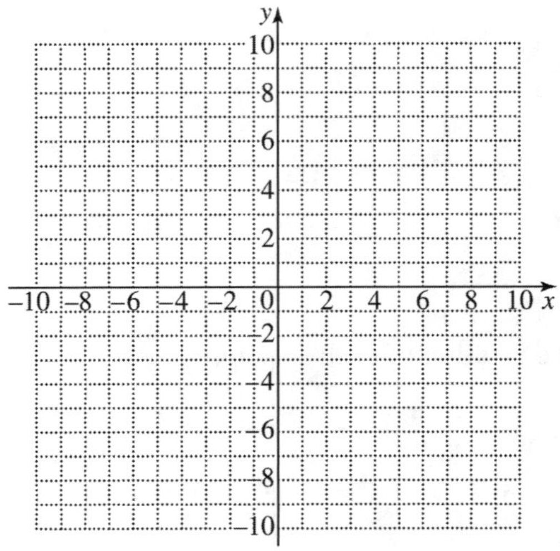

11. $\begin{cases} x^2 + y^2 < 9 \\ x^2 + y^2 > 1 \end{cases}$

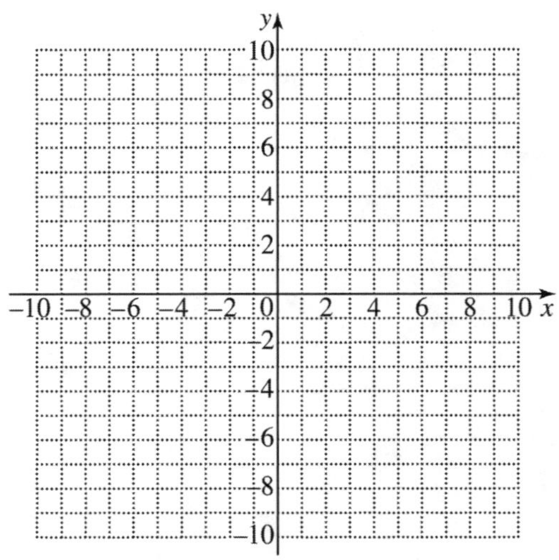

Name:
Instructor:

Date:
Section:

12. The cross section of the satellite antenna is a parabola given by the equation $y = \frac{1}{36}x^2$, with distances measured in feet. If the dish is 20 feet wide, how deep is it? Round your answer to one decimal point.

12. ______________________

Name:

Instructor:

Date:

Section:

Practice Test B

Sketch the graph of each equation.

1. $x^2 + y^2 = 25$

1. ______________________

a.

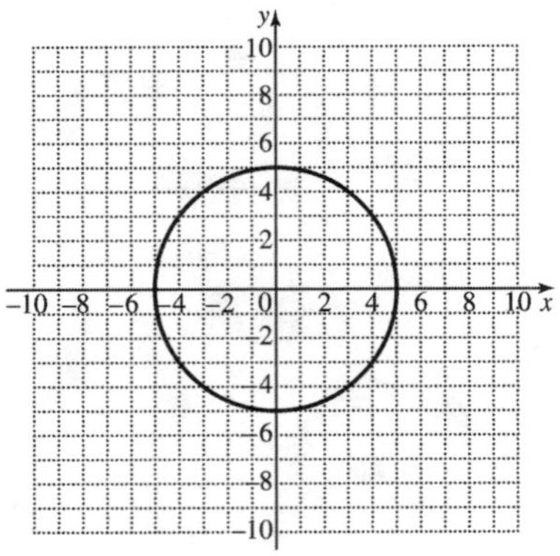

b.

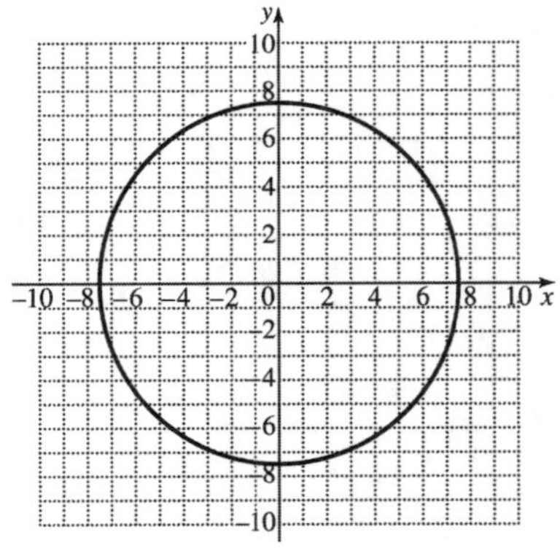

c.

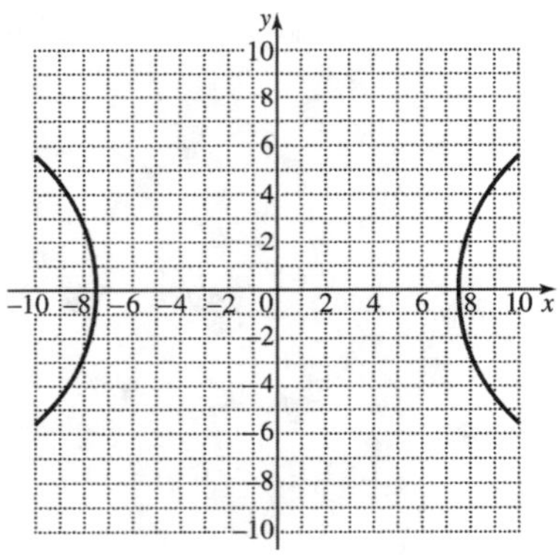

d.

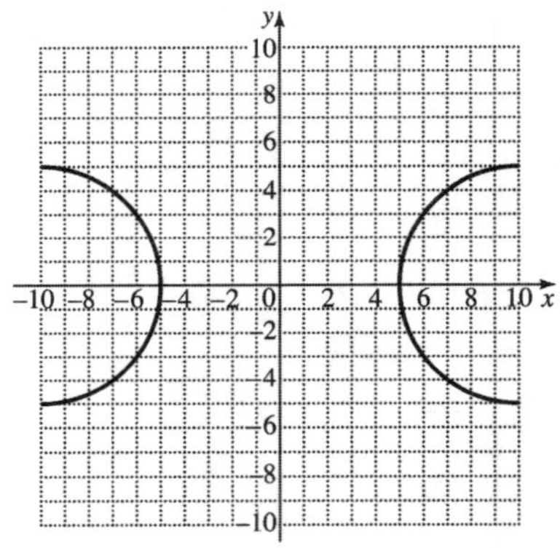

Name:
Instructor:

Date:
Section:

2. $\dfrac{y^2}{9} - x^2 = 1$

2. ____________________

a.

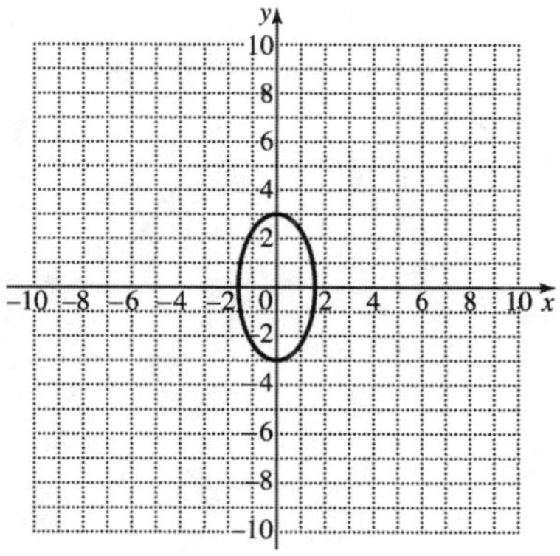

b.

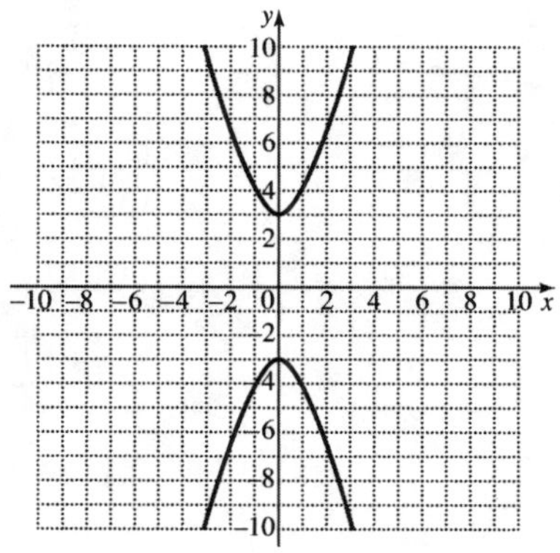

c.

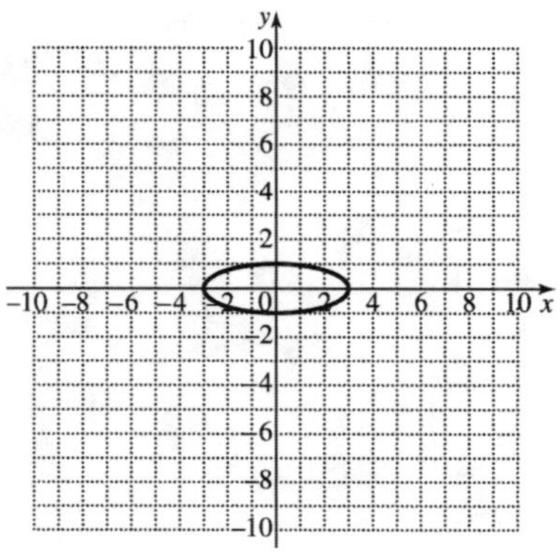

d.

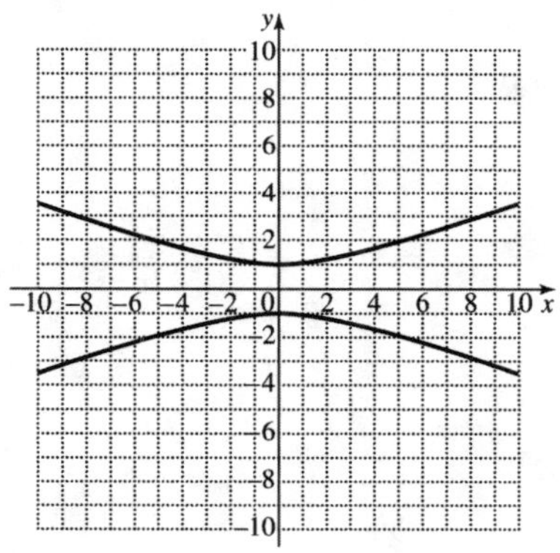

3. $x = (y-1)^2 - 3$

3. ____________________

a.

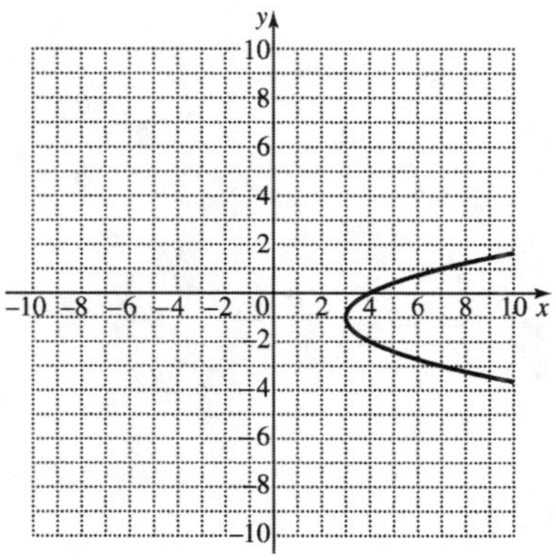

b.

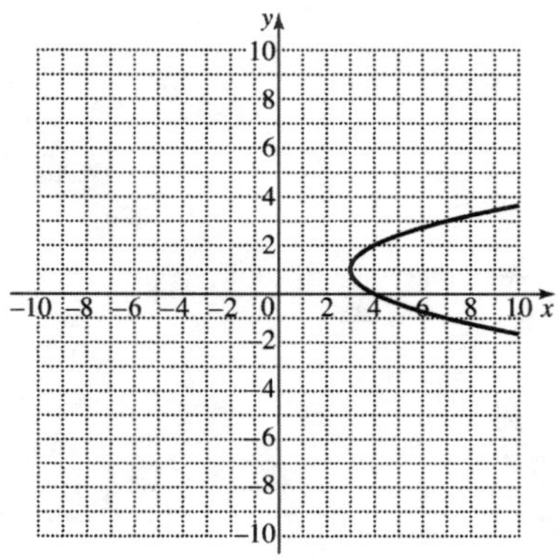

Name:
Instructor:

Date:
Section:

c.

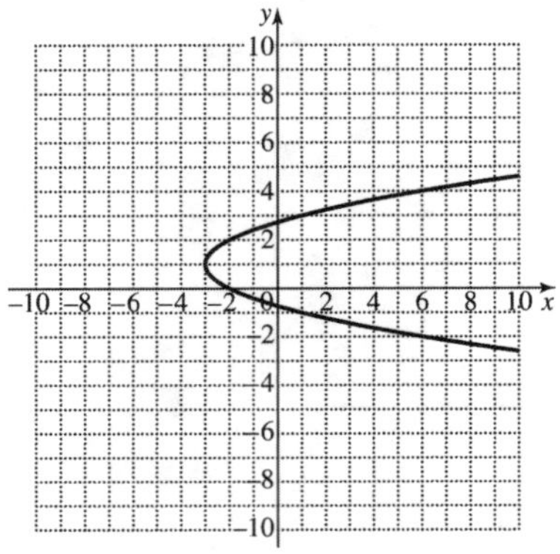

d.

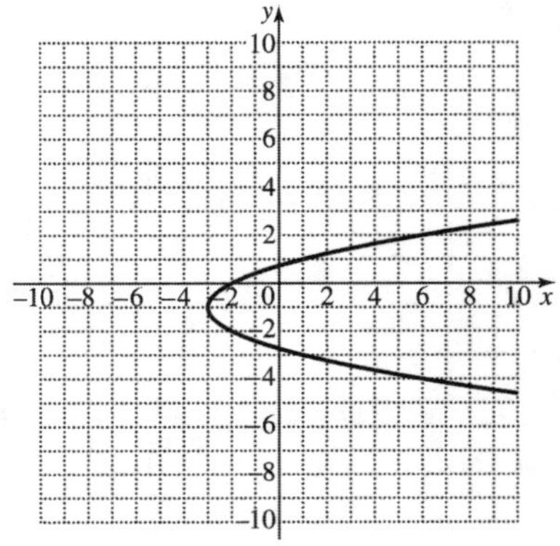

4. $\dfrac{(x-1)^2}{49}-\dfrac{(y+4)^2}{16}=1$

4. ______________________

a.

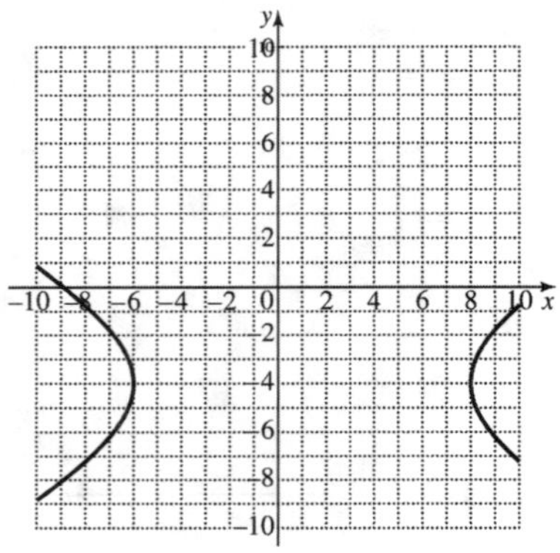

b.

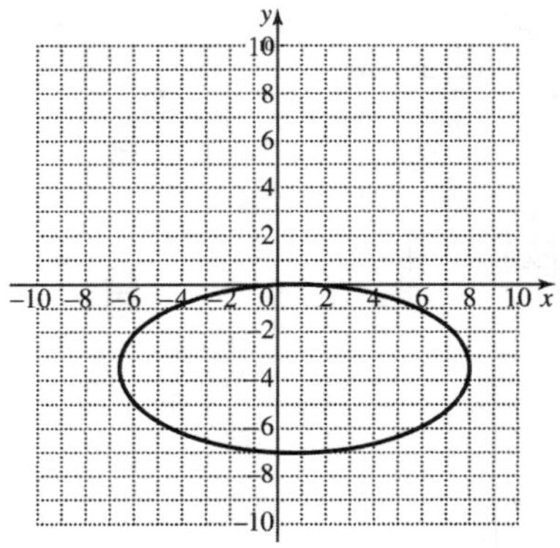

c.

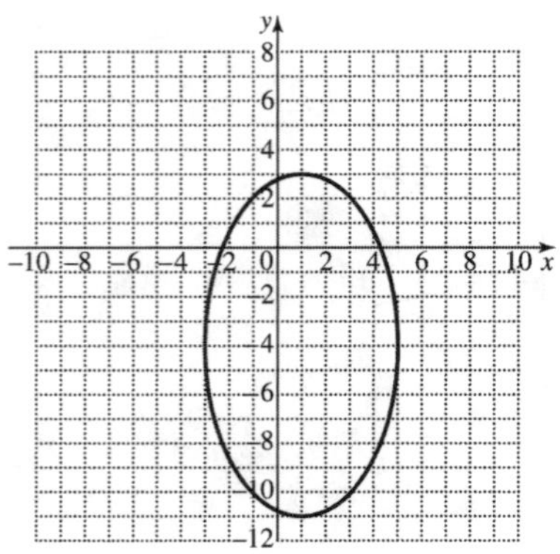

d.

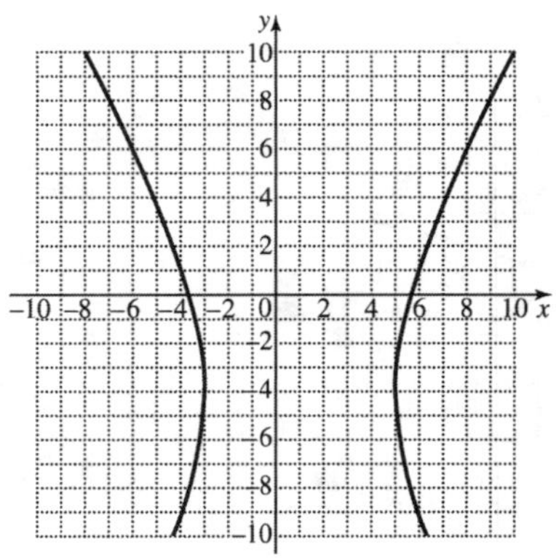

5. $x^2 + 16x + y^2 - 10x = 12$

5. ____________________

a.

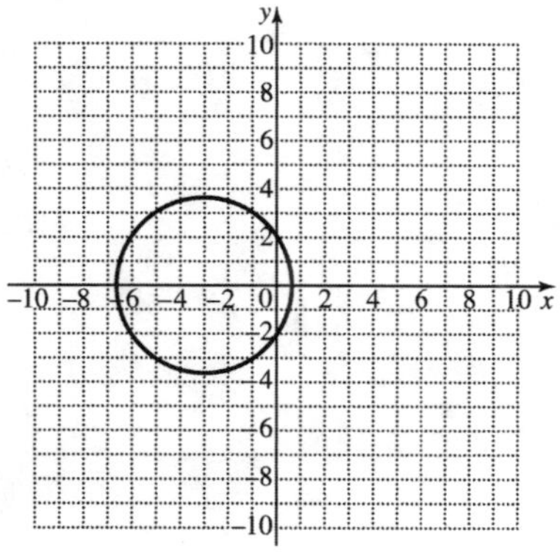

b.

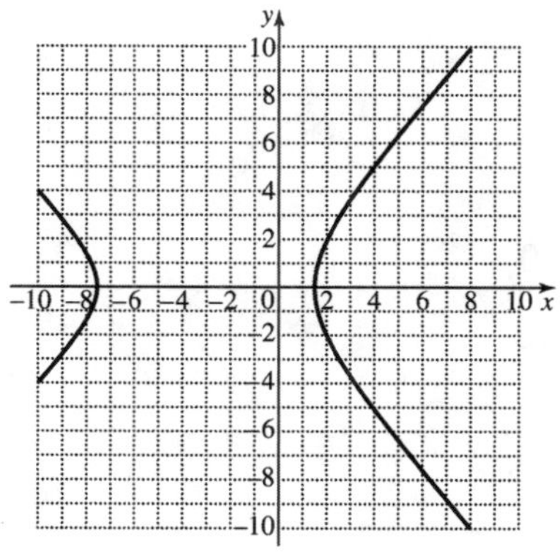

c.

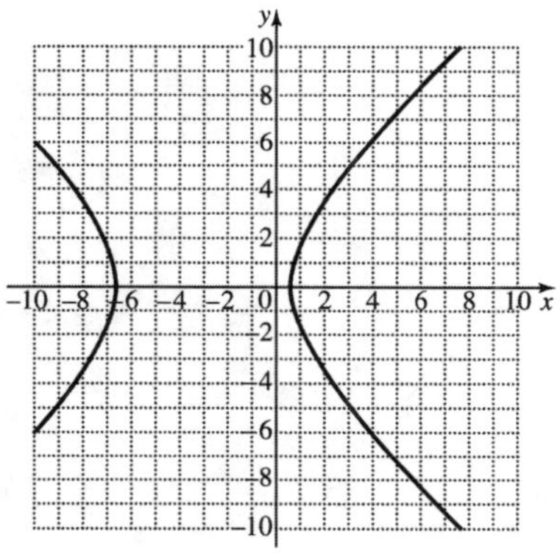

d.

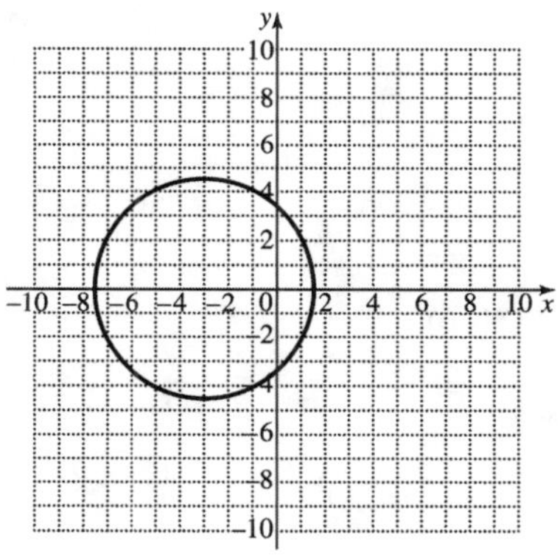

Solve each system.

6. $\begin{cases} x^2 + y^2 = 12 \\ -x^2 + y^2 = 8 \end{cases}$

6. ____________________

a. $(2,-10)(2,10)$

b. $(-\sqrt{2},\pm\sqrt{10})(\sqrt{2},\pm\sqrt{10})$

c. $(2,\pm 2\sqrt{2})(-2,\pm 2\sqrt{2})$

d. $\varnothing$

Name: **Date:**
Instructor: **Section:**

7. $\begin{cases} \dfrac{x^2}{4} - y^2 = 1 \\ 2x - y = 4 \end{cases}$

7. ____________________

a. $\left(4, \pm\sqrt{3}\right)$

b. $(2,0)\left(4, \sqrt{3}\right)$

c. $\left(\frac{34}{15}, \frac{8}{15}\right)(2,0)$

d. $\varnothing$

8. $\begin{cases} 2x^2 = y - 3 \\ x = -2y \end{cases}$

8. ____________________

a. $(1,5)(-2,1)$

b. $(-2,1)(2,1)$

c. $(4,-2)(2,-1)$

d. $\varnothing$

Graph the solution of each system.

9. $\begin{cases} x > y^2 - 4 \\ y < x^2 - 1 \end{cases}$

9. ____________________

a.

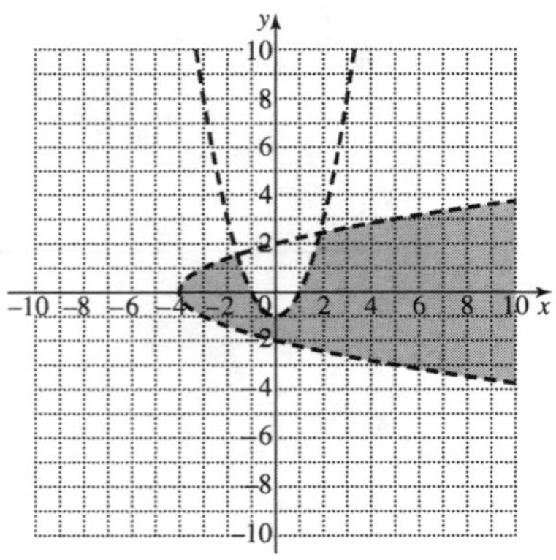

b.

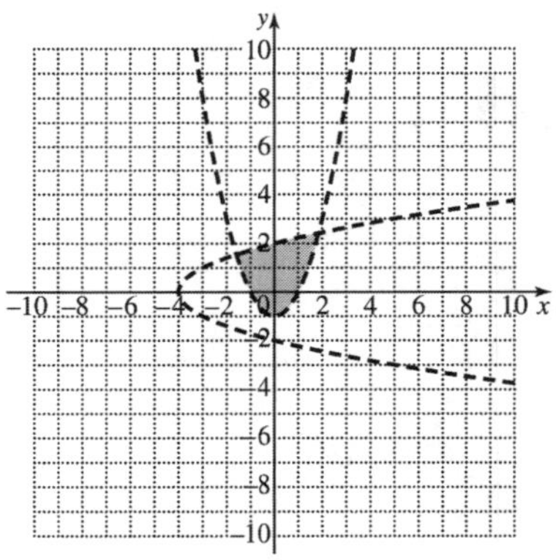

c.

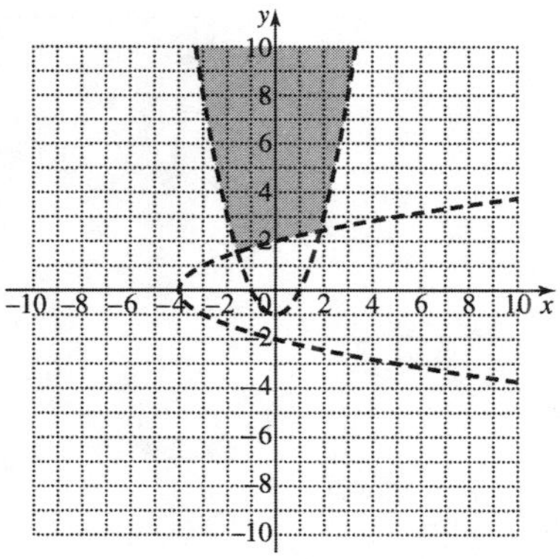

d.

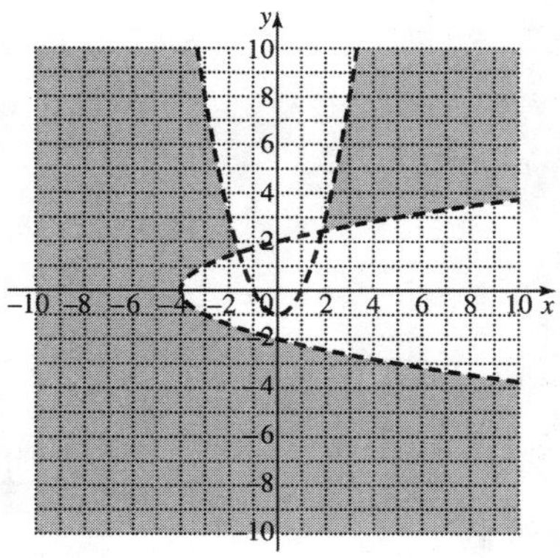

10. $\begin{cases} \dfrac{x^2}{4}+\dfrac{y^2}{9}<1 \\ x-2y \geq 6 \end{cases}$

10. ____________________

a.

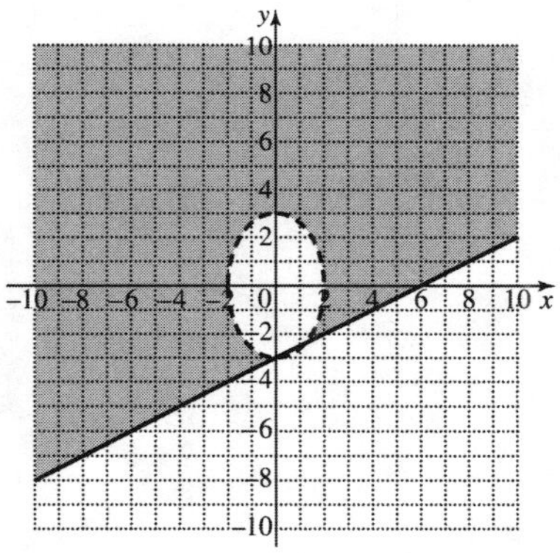

b.

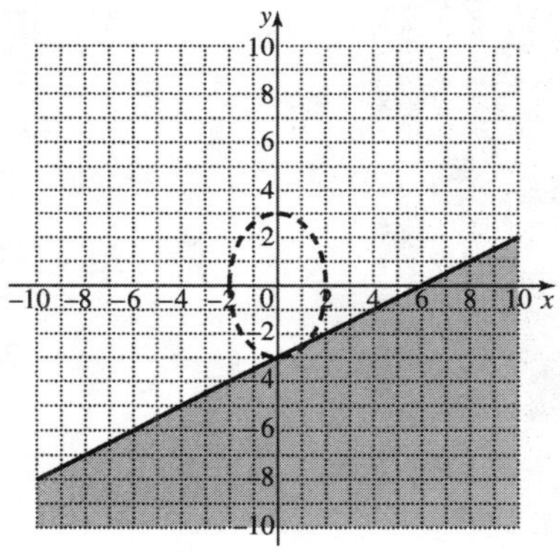

c.

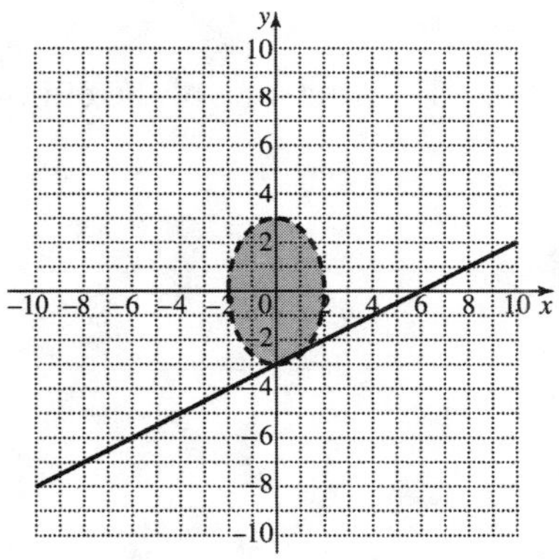

d.

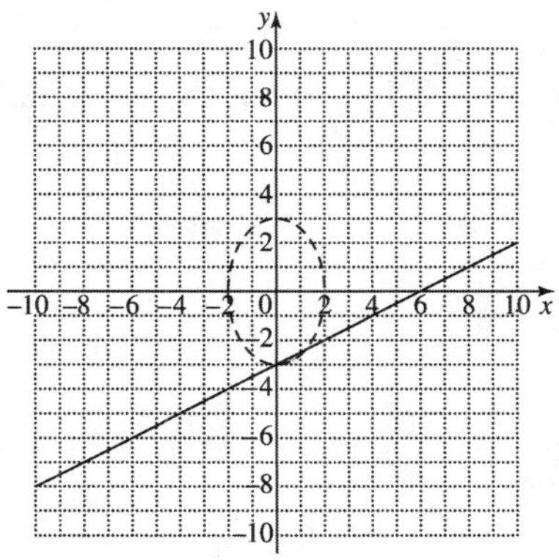

Name: **Date:**
Instructor: **Section:**

11. $\begin{cases} 4x^2 + y^2 < 64 \\ x^2 + y^2 \geq 1 \end{cases}$

11. ______________________

a.

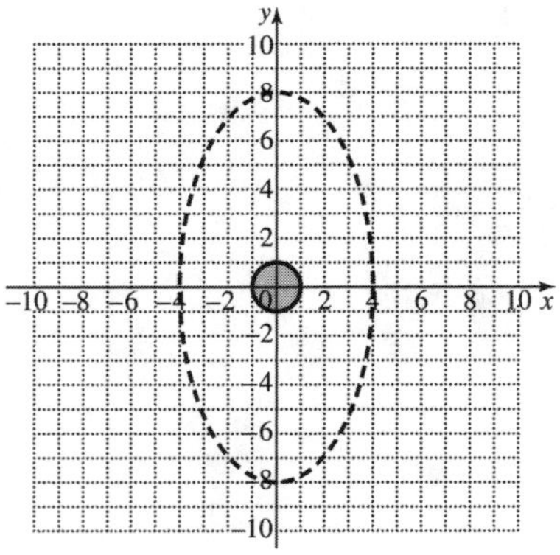

b.

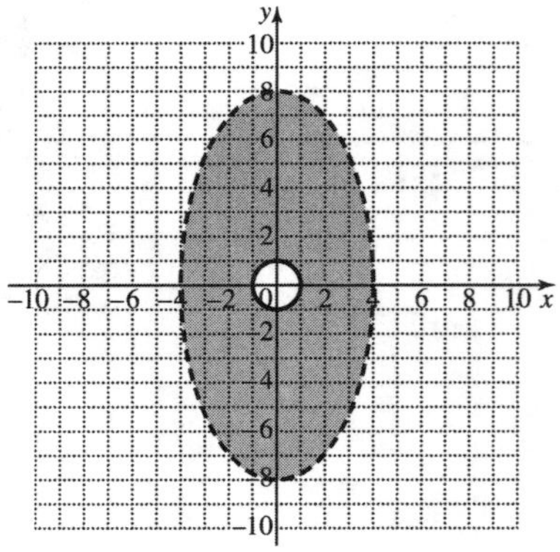

c.

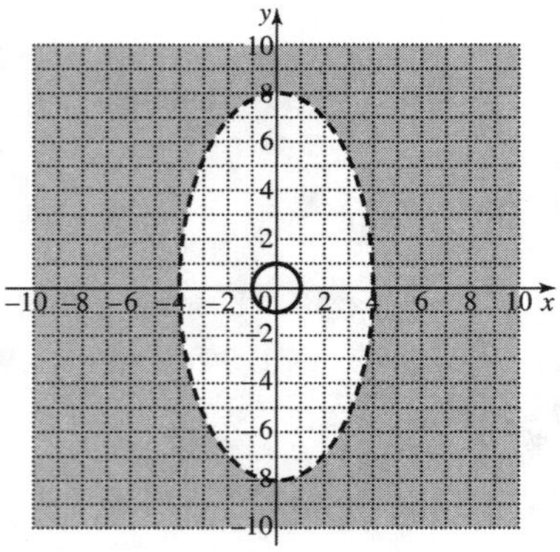

d.

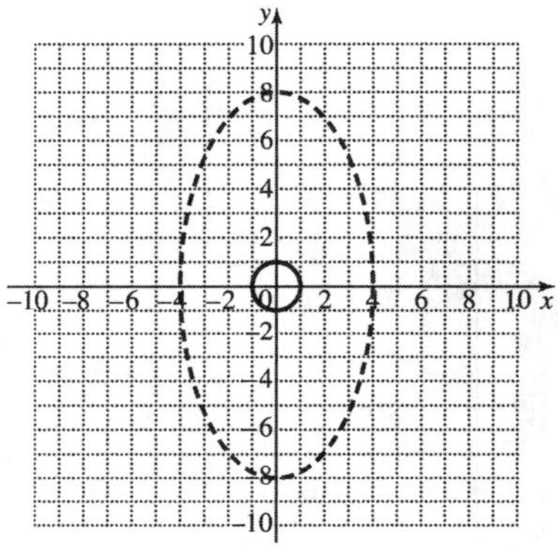

12. The area of a ellipse $\frac{x^2}{a^2} + \frac{y^2}{b^2} = 1$ is given by the equation $A = \pi ab$. Find the area of the ellipse $16x^2 + 9y^2 = 144$.

12. ______________________

a. 144π

b. 12π

c. 288π

d. 25π

Name:
Instructor:

Date:
Section:

Chapter 11 Sequences, Series, and the Binomial Theorem
Section 11.1 Sequences

Learning Objectives
1. Write the terms of a sequence given the general term.
2. Find the general term of a sequence.
3. Solve applications that involve sequences.

Vocabulary
Use the choices to complete each statement.

Finite **General Term** **Infinite**

1. A(n) ________________ sequence is a function whose domain is $\{1, 2, 3, \ldots\}$.
2. The ________________, or a_n, is the nth term of the sequence.
3. A(n) ________________ sequence is a function whose domain is $\{1, 2, 3, \ldots, n\}$

Objective 1

Write the first five terms of each sequence whose general term is given.

4. $a_n = n + 2$

4. ________________

5. $a_n = (-3)^{n-1}$

5. ________________

6. $a_n = n^2 - 4$

6. ________________

Name:
Instructor:

Date:
Section:

7. $a_n = \dfrac{1}{n+3}$

7. ____________________

8. $a_n = (-1)^n (2n)$

8. ____________________

Find the indicated term for each sequence whose general term is given.

9. $a_n = -2n;\ a_9$

9. ____________________

10. $a_n = \dfrac{n-4}{n+2};\ a_{10}$

10. ____________________

11. $a_n = \dfrac{(-1)^{n+1}}{n^2};\ a_{20}$

11. ____________________

Name:
Instructor:

Date:
Section:

12. $a_n = (n-3)(n+4);\ a_{34}$

12. ____________________

13. $a_n = 2^{3n-1};\ a_8$

13. ____________________

Objective 2

Find the general term a_n for each sequence whose first four terms are given.

14. $3, -9, 27, -81$

14. ____________________

15. $-2, -4, -8, -16$

15. ____________________

Name: **Date:**
Instructor: **Section:**

16. $\frac{1}{3}, \frac{1}{5}, \frac{1}{7}, \frac{1}{9}$

16. ____________________

17. $\frac{3}{5}, \frac{1}{2}, \frac{3}{7}, \frac{3}{8}$

17. ____________________

Objective 3

Solve.

18. Mrs. Laser agrees to give her son Mark an allowance of $0.10 on the first day of his 14-day vacation, $0.20 on the second day, $0.40 on the third day, and so on. Write an equation of a sequence whose terms correspond to Mark's allowance. Find the allowance Mark will receive on the last day of his vacation.

18. ____________________

19. George puts \$80 in his savings account. Each paycheck, he will put an additional \$40 more in the box. Find the amount in his account for the first five paychecks after the deposits.

19. ______________________

Concept Extension

Find the first five terms of each sequence. Round each term after the first to four decimal places.

20. $a_n = \dfrac{\sqrt{n+2}}{\sqrt{n}+2}$

20. ______________________

21. $a_n = \left(1 + \dfrac{0.025}{n}\right)^{n-1}$ Round to four decimal places of needed.

21. ______________________

Name: **Date:**
Instructor: **Section:**

Section 11.2 Arithmetic and Geometric Sequences

Learning Objectives
1. Identify arithmetic sequences and their common differences.
2. Identify geometric sequences and their common differences.

Vocabulary

Use the choices to complete each statement.

Arithmetic	**Difference**	**First**
Geometric	**Last**	**Ratio**

1. A(n) ______________________ sequence is one in which each term (after the first) differs from the term before it by a constant common _________________.
2. A(n) ______________________ sequence is one in which each term (after the first) is derived from multiplying the proceeding term by a constant common ___________________.
3. In a sequence the term a_1 is the ______________________ term.

Objective 1

Write the first five terms of the arithmetic or geometric sequence whose first term a_1, and common difference, d are given.

4. $a_1 = 6;\ d = -2$ 4. ______________________

5. $a_1 = 25;\ d = 7$ 5. ______________________

6. $a_1 = -2;\ d = -5$ 6. ______________________

Name:
Instructor:

Date:
Section:

7. $a_1 = 10;\ d = 16$

7. ____________________

Find the indicated term of each sequence.

8. The ninth term of the arithmetic sequence whose first term is 13 and whose common difference is 9.

8. ____________________

9. The eleventh term of the arithmetic sequence whose first term is 4 and whose common difference is 3.

9. ____________________

10. The sixth term of the arithmetic sequence whose first term is 1 and whose common difference is – 6.

10. ____________________

11. The twenty-fifth term of the arithmetic sequence 20, 18, 16,…

11. ____________________

12. The tenth term of the arithmetic sequence 12, 8, 4, …

12. ____________________

Name: **Date:**
Instructor: **Section:**

Objective 2

Write the first five terms of the geometric sequence whose first term a_1 , and common ratio, r are given.

13. $a_1 = 48;\ r = \frac{1}{2}$

13. ____________________

14. $a_1 = 10;\ \ r = -3$

14. ____________________

15. $a_1 = -1;\ \ r = 4$

15. ____________________

Find the indicated term of each sequence.

16. The fifth term of the geometric sequence whose first term is 8 and whose common ratio is $\frac{1}{4}$.

16. ____________________

17. The seventh term of the geometric sequence whose first terms in – 2 and whose common ratio is 3.

17. ____________________

Name: **Date:**
Instructor: **Section:**

18. The sixth term of the geometric sequence $3, 12, 48 \ldots$

18. ____________________

19. The eighth term of the geometric sequence $4, -2, 1 ..$

19. ____________________

Concept Extension

Write the first four terms of the following sequences.

20. $a_1 = 2560.25$; $d = -1450.75$

20. ____________________

21. $a_1 = 32.3$; $r = -2.2$

21. ____________________

Name: **Date:**
Instructor: **Section:**

Section 11.3 Series

Learning Objective

1. Identify finite and infinite series and use summation notation.
2. Find partial sums.

Objective 1

Evaluate.

1. $\sum_{i=4}^{7} 2i+4$ 1. ____________________

2. $\sum_{i=3}^{8} 4i$ 2. ____________________

3. $\sum_{i=6}^{10} i(i-2)$ 3. ____________________

Name: **Date:**
Instructor: **Section:**

4. $\sum_{i=2}^{8} \frac{2i}{i+4}$

4. ____________________

5. $\sum_{i=2}^{6} 2^{i+2}$

5. ____________________

Write each series with summation notation.

6. $4+9+16+25+36+49$

6. ____________________

7. $5+7+9+11+13$

7. ____________________

8. $6+11+18+27+38$

8. ____________________

Name: **Date:**
Instructor: **Section:**

Objective 2

Find each partial sum.

9. Find the sum of the first two terms of the sequence whose general term is $a_n = n(n-6)$.

9. ____________________

10. Find the sum of the first four terms of the sequence whose general term is $a_n = (-1)^{2n+1}$.

10. ____________________

11. Find the sum of the fist six terms of the sequence whose general term is $a_n = n(n-3)$.

11. ____________________

12. Find the sum of the first five terms of the sequence whose general term is $a_n = -\frac{n}{2}$.

12. ____________________

Name:
Instructor:

Date:
Section:

Concept Extension

13. Show that $\sum_{i=1}^{5} \frac{i^2}{i} \neq \frac{\sum_{i=1}^{5} i^2}{\sum_{i=1}^{5} i}$. What do you think $\sum_{i=1}^{5} \frac{i^2}{i}$ might equal? Show that it does.

13. ____________________

Name: **Date:**
Instructor: **Section:**

Section 11.4 Partial Sums of Arithmetic and Geometric Sequences

Learning Objectives
1. Find the partial sum of an arithmetic sequence.
2. Find the partial sum of a geometric sequence.
3. Find the sum of the terms of an infinite geometric sequence.

Objective 1

Use the partial sum formula to find the partial sum of the given arithmetic and geometric sequence.

1. Find the sum of the first six terms of the arithmetic sequence $3, 6, 9, \ldots$

1. ______________________

2. Find the sum of the first five positive even integers.

2. ______________________

3. Find the sum of the first seven terms of $-15, -8, -1 \ldots$

3. ______________________

4. Find the sum of the first four positive odd integers.

4. ______________________

Name:
Instructor:

Date:
Section:

Objective 2

5. Find the sum of the first five terms of the geometric sequence $4, 12, 36, \ldots$

5. ____________________

6. Find the sum of the first six terms of the geometric sequence $32, -16, 8, \ldots$

6. ____________________

7. Find the sum of the first four terms of the geometric sequence $1, \frac{1}{4}, \frac{1}{16}, \ldots$

7. ____________________

8. Find the sum of the first five terms of the geometric sequence if $a_1 = 16$ and $r = -\frac{2}{3}$.

8. ____________________

Name: **Date:**
Instructor: **Section:**

Objective 3

Find the sum of the terms of each infinite geometric sequence.

9. $32, 16, 8 \ldots$

9. ____________________

10. $-10, -5, -\frac{5}{2}$

10. ____________________

11. $4, -\frac{1}{2}, \frac{1}{16}, \ldots$

11. ____________________

12. $3, \frac{1}{3}, \frac{1}{27}, \ldots$

12, ____________________

Name:
Instructor:

Date:
Section:

Concept Extension

13. Write $0.44\overline{4}$ as an infinite geometric series and use the formula for S_∞ to write it as a rational number.

13. ____________________

14. Write $0.36\overline{36}$ as an infinite geometric series and use the formula for S_∞ to write it as a rational number.

14. ____________________

Name: **Date:**
Instructor: **Section:**

Section 11.5 The Binomial Theorem

Learning Objectives

1. Use Pascal's triangle to expand binomials.
2. Evaluate factorials.
3. Use the binomial theorem to expand binomials.
4. Find the *n*th term in the expansion of a binomial raised to a positive power.

Objective 1

Use Pascal's triangle to expand the binomial.

1. $(m+n)^3$ — 1. ______________________

2. $(a-b)^4$ — 2. ______________________

3. $(x+y)^8$ — 3. ______________________

Objective 2

Evaluate each expression.

4. $\dfrac{8!}{4!}$ — 4. ______________________

Name:
Instructor:

Date:
Section:

5. $\dfrac{24!}{20!}$

5. ________________

6. $\dfrac{16!}{7!9!}$

6. ________________

Objective 3

Use the binomial formula to expand each binomial.

7. $(x-y)^5$

7. ________________

8. $(2x+3)^6$

8. ________________

Name: **Date:**
Instructor: **Section:**

9. $(5a-2b)^4$ 9. ____________

10. $(4+3x)^6$ 10. ____________

Objective 4

Find the indicated term.

11. The fifth term of the expansion of $(2x+5z)^8$. 11. ____________

12. The fourth term of the expansion of $(2r-s)^5$. 12. ____________

Name: **Date:**
Instructor: **Section:**

13. The third term of the expansion of $(m-3n)^7$. 13. ____________________

Concept Extension

14. Expand the expression $(\sqrt{a}-\sqrt{6})^4$. 14. ____________________

15. Find the term containing x^3 in the expansion of $(\sqrt{x}+\sqrt{2})^{10}$. 15. ____________________

Name: **Date:**
Instructor: **Section:**

Chapter 11 Vocabulary

Vocabulary Word	**Definition**	**Example**
Infinite Series	A function whose domain is the set of natural numbers $\{1,2,3,4...\}$	$1,3,5,7,....$
Finite Series	A function whose domain is the set of natural numbers $\{1,2,3,...,n\}$	$2,4,6,8,10$
Arithmetic Sequence	A sequence in which each term differ from the previous term by a common difference.	$3,6,9,12,15,...$
General term of an Arithmetic Sequence.	$a_n = a_1 + (n-1)d$	$a_n = 5 + (n-1)6$
Geometric Sequence	A sequence in which each term (after the first term) is found by multiplying the previous term by a common ratio	$8,-16,32,-64,...$
General Term of a Geometric Sequence.	$a_n = a_1 r^{n-1}$	$a_n = 3(4)^{n-1}$
Series	Sum of terms in a sequence.	1+3+5+7
Summation Notation	$\sum$: sigma	$\sum_{i=2}^{6} i = 2+3+4+5+6$
Partial Sum	S_n	S_3 of 1,2,3 $1+2+3=6$

Name:
Instructor:

Date:
Section:

Partial Sum of Arithmetic Sequence	$S_n = \frac{n}{2}(a_1 + a_n)$	3,5,7,9.... $S_3 = \frac{3}{2}(3+7) = 15$
Partial Sum of a Geometric Sequence	$S_n = \frac{a_1(1-r^n)}{1-r}$	2,4,8,16... $S_3 = \frac{2(1-2^3)}{(1-2)} = 14$
Sum of an Infinite Geometric Sequence	$S_\infty = \frac{a_1}{1-r}$	4,2,1,.... $S_\infty = \frac{4}{1-\frac{1}{2}} = 8$

Name:
Instructor:

Date:
Section:

Practice Test A

Find the indicated term(s) of the given sequences.

1. The first six terms of the sequence $a_n = \frac{(2n+1)}{n+4}$.

1. ______________________

2. The ninth term of the sequence $a_n = 5\left(\frac{1}{3}\right)^{n-1}$.

2. ______________________

3. The general term of the sequence $7, -1, \frac{1}{7}, \ldots$

3. ______________________

4. The general term of the sequence $24, 10, -4, \ldots$

4. ______________________

Name:
Instructor:

Date:
Section:

Find the partial sum of the given sequence.

5. S_7 of the sequence $a_n = 5 + (n-1)6$

5. ______________________

6. S_6 of the sequence $a_n = -4\left(-\frac{1}{2}\right)^{n-1}$

6. ______________________

7. S_∞ of the sequence $a_1 = 144$ and $r = \frac{1}{6}$

7. ______________________

8. S_4 of the sequence $8, -16, 32, \ldots$

8. ______________________

Name:
Instructor:

Date:
Section:

9. $\sum_{i=2}^{7} i^2 - 3i + 5$

9. ____________________

10. $\sum_{i=1}^{6} 6(3)^i(-1)^i$

10. ____________________

Expand each binomial.

11. $(2x - y)^4$

11. ____________________

12. $(a + 3b)^5$

12. ____________________

Name:
Instructor:

Date:
Section:

Practice Test B

Find the indicated term(s) of the given sequences.

1. The first six terms of the sequence $a_n = \dfrac{n^2 - 2}{n + 6}$.

1. ____________________

a. $7, 8, 9, 10, 11, 12$

b. $1, 2, 7, 14, 23, 34$

c. $-\frac{5}{7}, \frac{3}{4}, \frac{11}{9}, \frac{9}{5}, \frac{27}{11}, \frac{19}{6}$

d. $-\frac{1}{7}, \frac{1}{4}, \frac{7}{9}, \frac{7}{5}, \frac{23}{11}, \frac{17}{6}$

2. The ninth term of the sequence $a_n = 16\left(\dfrac{1}{2}\right)^{n-1}$.

2. ____________________

a. $1/16$

b. $1/32$

c. 8192

d. 4096

3. The general term of the sequence $18, -9, \frac{9}{2}, \ldots$

3. ____________________

a. $a_n = 18\left(\dfrac{1}{2}\right)^n$

b. $a_n = 18\left(-\dfrac{1}{2}\right)^{n-1}$

c. $a_n = -18\left(\dfrac{1}{2}\right)^{n-1}$

d. $a_n = 18 - 9(n - 1)$

Name: **Date:**
Instructor: **Section:**

4. The general term of the sequence $15, -5, -25, \ldots$

4. ____________________

a. $a_n = 15 + (n-1)(-20)$

b. $a_n = 15\left(\frac{1}{3}\right)^n$

c. $a_n = 15 + (n-1)20$

d. $a_n = 15\left(-\frac{1}{3}\right)^{n-1}$

Find the partial sum of the given sequence.

5. S_7 of the sequence $a_n = 22 + (n-1)(-8)$

5. ____________________

a. – 12

b. – 22

c. – 14

d. – 18

6. S_6 of the sequence $a_n = 100\left(\frac{1}{5}\right)^{n-1}$

6. ____________________

a. 613

b. $\frac{15624}{125}$

c. $\frac{16130}{26}$

d.. 625

Name: **Date:**
Instructor: **Section:**

7. S_{∞} of the sequence $a_1 = 150$ and $r = \frac{1}{5}$

7. ____________________

a. 184

b. 191

c. $\frac{401}{2}$

d. $\frac{375}{2}$

8. S_{∞} of the sequence $340, 170, 85, \ldots$

8. ____________________

a. 600

b. 680

c. 595

d. 740

9. $\sum_{i=4}^{9} 3i + 6$

9. ____________________

a. 156

b. 160

c. 153

d. 147

Name: **Date:**
Instructor: **Section:**

10. $\sum_{i=3}^{9} i(2i+3)$ 10. ______________

a. 686 b. 724

c. 622 d. 580

11. Find the third term of the expansion of $(4x+y)^5$. 11. ______________

a. $160x^2y^3$ b. $1280x^4y$

c. $20xy^4$ d. $640x^3y^2$

12. Find the sixth term of the expansion of $(c-2d)^8$. 12. ______________

a. $1120c^4d^4$ b. $-1792c^3d^5$

c. c^6 d. $1792c^2d^6$

Chapter 1
Section 1.2

1. Natural numbers
3. Rational number
5. Integers
7. Whole numbers.
9. – 6
11. 103
13. 0
15. $\not\in$
17. – 2
19. – 5.67
21. 67
23. 12 – 3x
25. True

Section 1.3

1.40
3. $-\frac{7}{15}$
5. $-\frac{21}{10}$
7. – 49
9. – 0.00001
11. – 2
13. 38
15. 316
17. – 31
19. – 11
21. $\frac{1}{24}$

Section 1.4

1. $2x+5=-14$
3. $\frac{r}{-9}\neq 30$
5. 0; undefined
7. $4; -\frac{1}{4}$
9. $3x$
11. $(8x+7y)+3z$
15. $112-x$
17. $19.95n$
19. Answers will vary.

Practice Test A

1. – 21
2. $\frac{69}{4}$
3. 30
4. – 8
5. $\frac{3}{8}$
6. 10(q + (-8))
7. $7-\sqrt{n}\geq 9$
8. $|3t|\neq \frac{1}{2}t+8$
9. Additive inverse property
10. Associative property of addition
11. Commutative property of addition.
12. A = 10b
13. $\frac{11}{12}x-\frac{37}{24}$
14. c – 22
15. – 1.2x – 1.1
16. >
17. =
18. {2, 4, 6, 8, 10}
19. –25
20. Not a real number

Practice Test B.

1. C
2. A
3. A
4. B
5. C
6. A
7. B
8. C
9. B
10. B
11. B
12. A
13. B
14. D
15. A
16. D
17. B
18. A
19. C
20. B

Chapter 2
Section 2.1

1. Solution
3. Multiplication
5. – 3

7. – 21
9. $\frac{22}{3}$
11. 10
13. – 12
15. $\frac{2}{3}$
17. – 8
19. All real numbers
21. All real numbers
23. 84

Section 2.2

1. $x^2 - 12x$
3. 65x + 30 cents
5. $\pi x + 4\pi$
7. 24, 25, 26
9. 160 females
11. $560.49
13. 20 cc of 30% solution

Section 2.3

1. $\frac{y-b}{m} = x$
3. $\frac{5}{9}(F-32) = C$
5. $\frac{25}{n} - a = L$
7. $\frac{2(S-B)}{p} = l$
9. 4 inches
11. $586.85
13. 45 m by 55 m
15. 74.07 kg

Section 2.4

1. $(3,\infty)$
3. $(-2,1.5]$
5. $(-\infty,-6)$
7. $(-\infty,9]$
9. $[-3,\infty)$
11. $(-13,\infty)$
13. $[\frac{79}{3},\infty)$
15. $\{x \mid x \leq 1040\}$
17. Answers may vary

Section 2.5

1. Compound
3. Union; $\cup$
5. $\{1,15,19,23,29\}$
7. $\varnothing$
9. $[-4,8]$
11. $[3,30]$
13. $[-3,\frac{3}{2}]$
15. $\{x \mid x$ is an odd number or $x = 2$ or $x = 12\}$
17. $(-\infty,2)$
19. $(-4,\infty)$
21. $(-\infty,1)\cup(5,\infty)$
23. $(-\frac{4}{7},\frac{10}{3}]$

Section 2.6

1. 8, - 8
3. 10, - 10
5. 3, - 2
7. 8, 4
9. 2
11. $\varnothing$
12. 5, 1
13. $|x+5| = 4$

Section 2.7

1. $(-5,5)$
3. $(-3,9)$
5. $[-16,14]$
7. $\varnothing$
9. $[-1,8]$
11. $(-\infty,-4]\cup[6,\infty)$
13. $(-\infty-21)\cup(27,\infty)$
15. $(-\infty,-2]\cup[8,\infty)$

Practice Test A

1. – 16
2. $17/2$
3. $2/5$
4. ∅
5. 7, - 3
6. ∅
7. $148/7$
8. $7/5, -9$
9. 6
10. $L = \dfrac{P - 2w}{2}$
11. $r = \dfrac{T}{s^2 - ps}$
12. $b = \dfrac{2A}{h}$
13. $(-2, \infty)$
14. $(-11/2, \infty)$
15. $(-\infty - 7]$
16. $(-\infty, 15)$
17. $[2, 8]$
18. $(-2, 5)$
19. 21, 23, 25, and 27
20. {16, 18, 20, 22, 24}

Practice Test B

1. B
2. C
3. D
4. A
5. B
6. C
7. A
8. A
9. B
10. D
11. A
12. C
13. C
14. D
15. B
16. A
17. D
18. B
19. A
20. D

Chapter 3
Section 3.1

1. 10 to the left, 16 up; II
3. 7 to the right, 19 up; I
5. No
7. No
9.

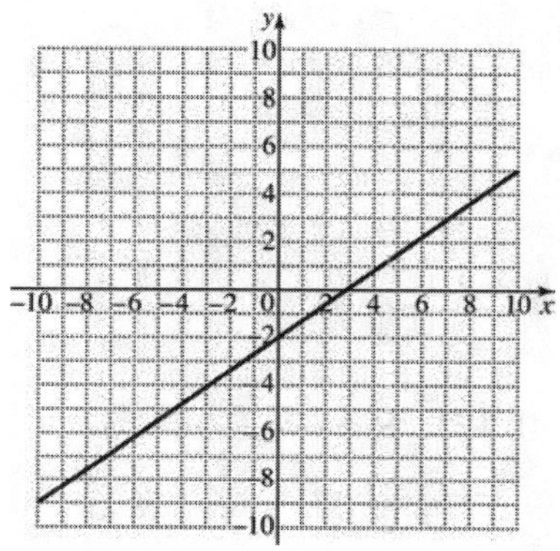

11.

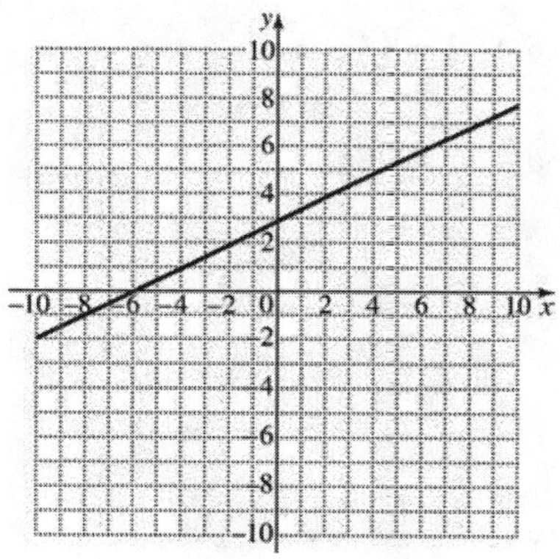

13.

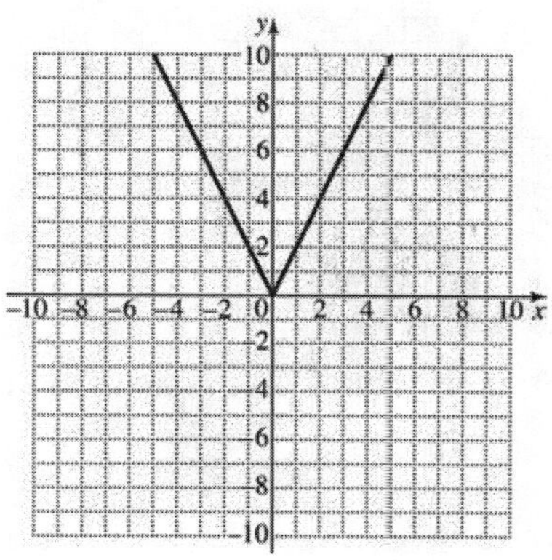

15.

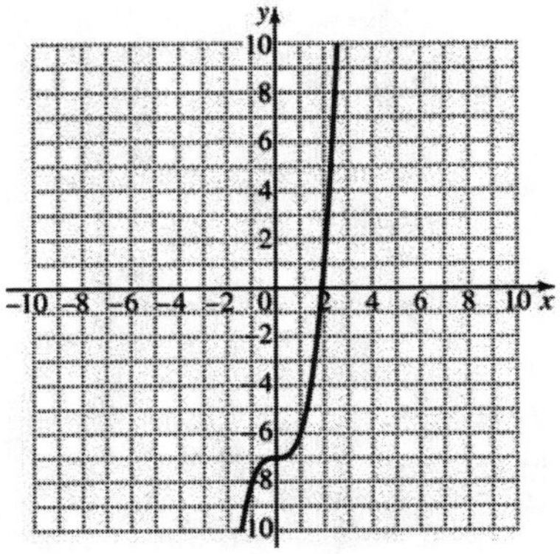

17. The second graph shifts 2 units to the right, and 5 units up.

Section 3.2

1. Function
3. y
5. Relation
7. x
9. D = {1, -7, 0, 12}; R = {7, 3, -6, 15}
11. Yes
13. Yes
15. Yes
17. No
19. D= $(-\infty, \infty)$; R = $(-\infty, \infty)$
21. 24
23. 3a – 4; 3a – 3h – 4

Section 3.3

1. Horizontal
3. Vertical
5.

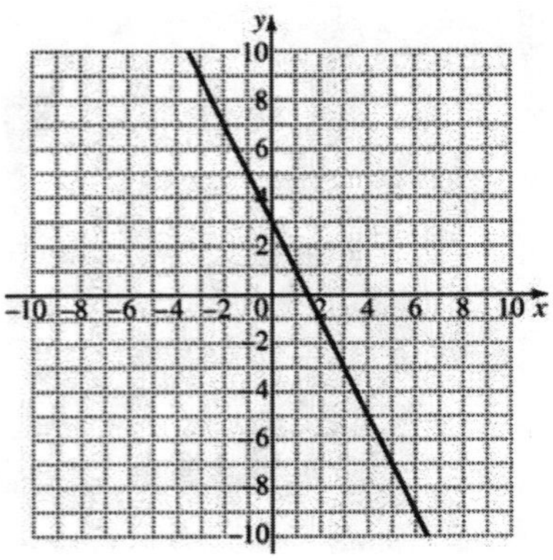

7.

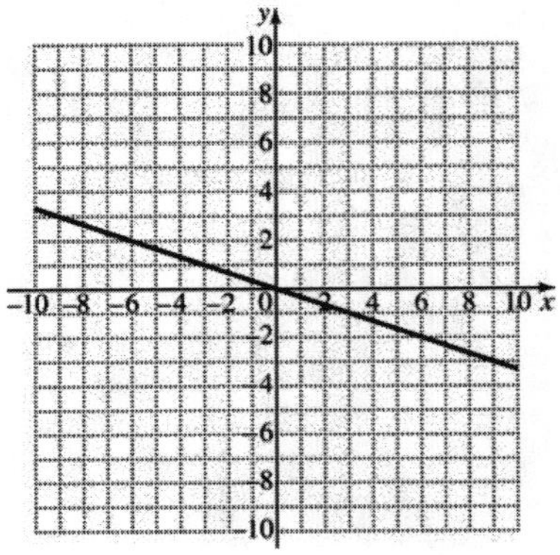

9.

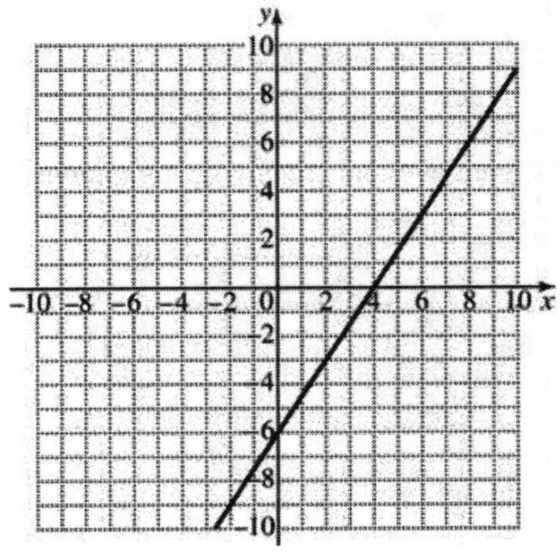

11.

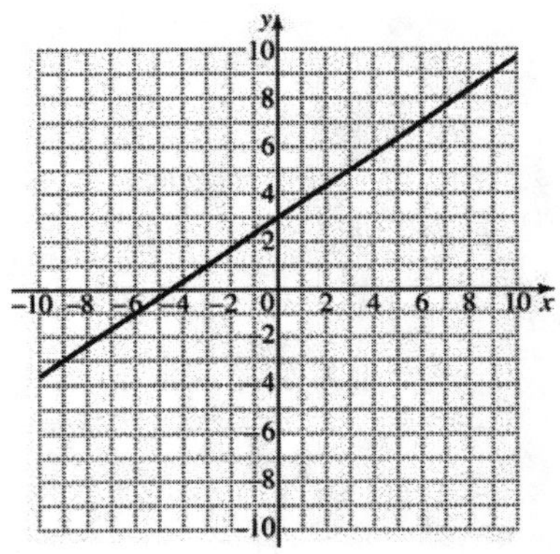

13.

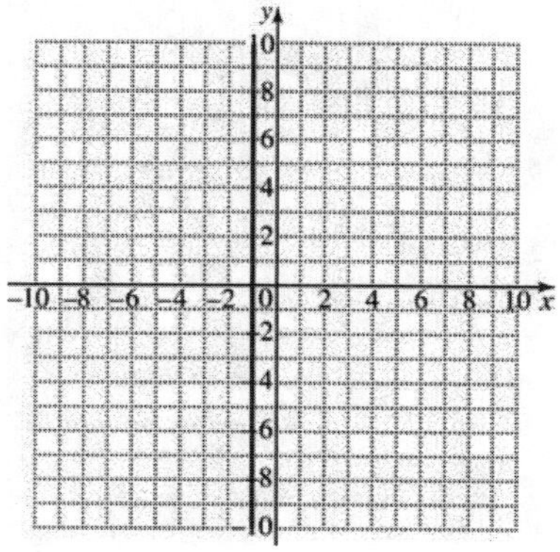

15.

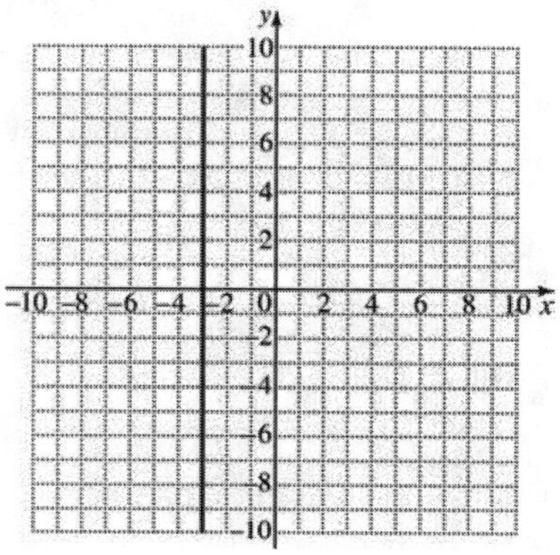

17. coefficient = 0; slope = 0

Section 3.4

1. Equal; Different
3. Slope
5. Slope-Intercept
7. $\frac{1}{3}$
9. undefined
11. $\frac{4}{7}$

13a. m = 33; y-intercept (0, 42)
13b. The number of WiFi hotspots increase by 33 thousand every year.
13c. There were 42 thousand hotspots in 2003.

15. undefined
17. undefined
19. Neither
21. Perpendicular
23. $\frac{4}{5}$

Section 3.5

1. $y = -2x - 4$
3. $y = -\frac{3}{5}x - \frac{7}{9}$
5.

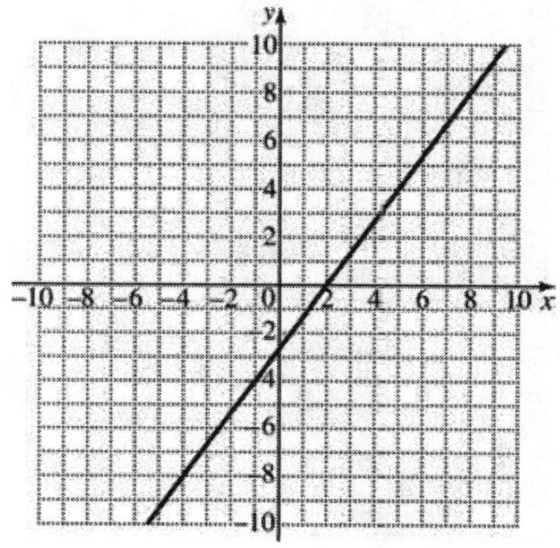

7.

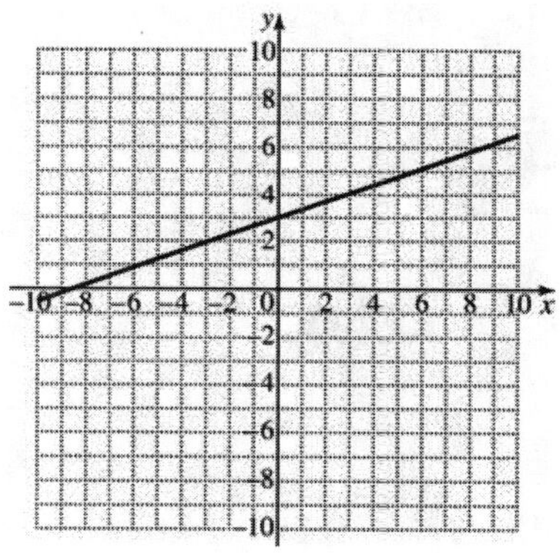

9. $y = \frac{1}{4}x - 5$
11. $y = -7x + \frac{41}{10}$
13. $x = 2$
15. $y = -\frac{1}{3}$
17. $f(x) = \frac{1}{3}x - \frac{10}{3}$
19. $f(x) = -\frac{3}{2}x - 23$
21. $f(x) = -\frac{13}{19}x + \frac{43}{380}$

Section 3.6

1.

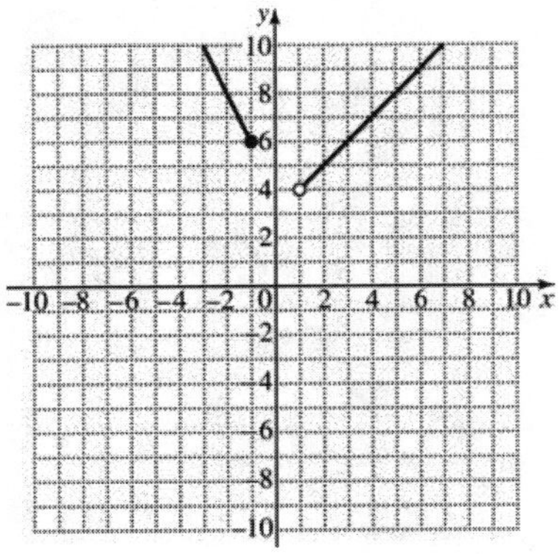

3.

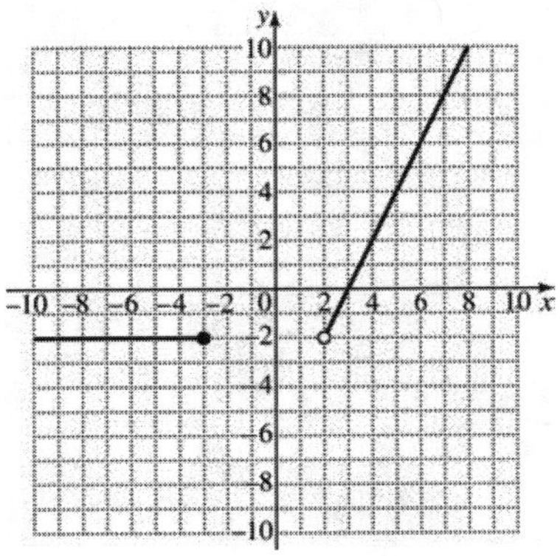

5.

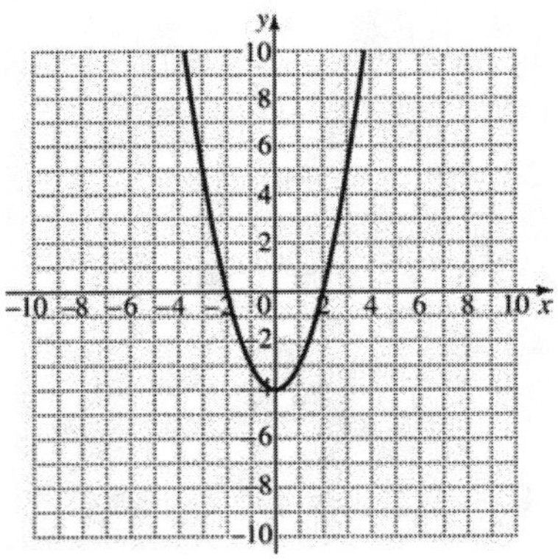

7.

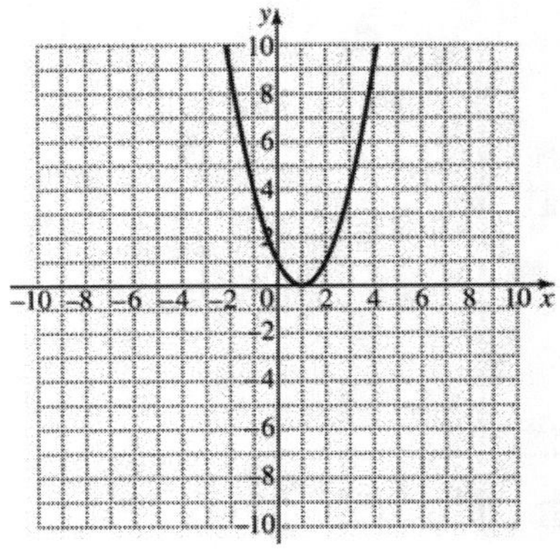

9.

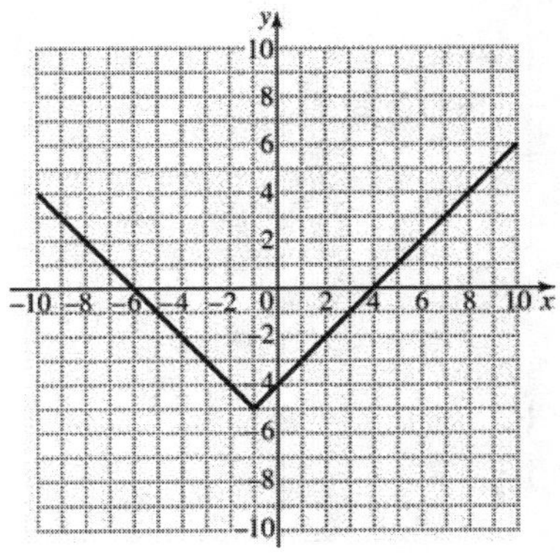

11.

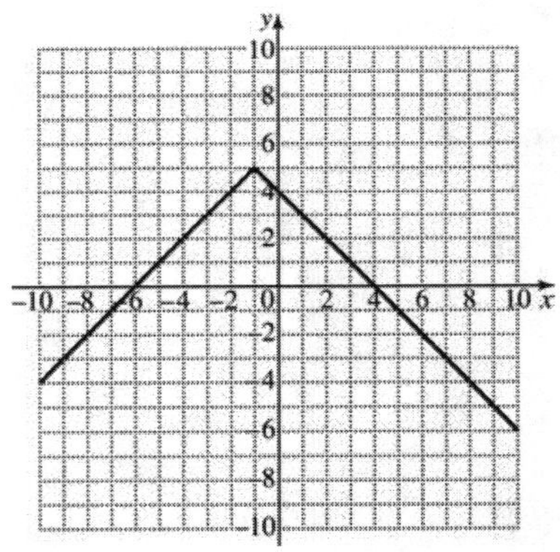

13.

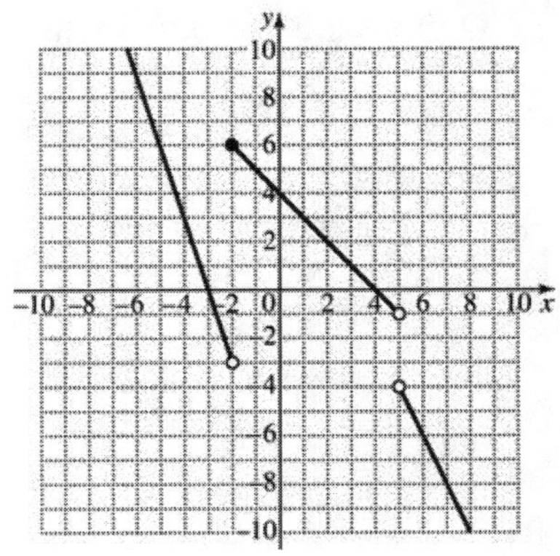

Section 3.7

1.

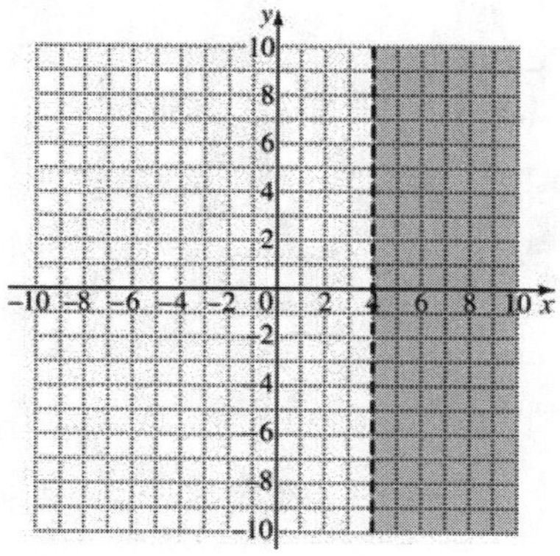

3.

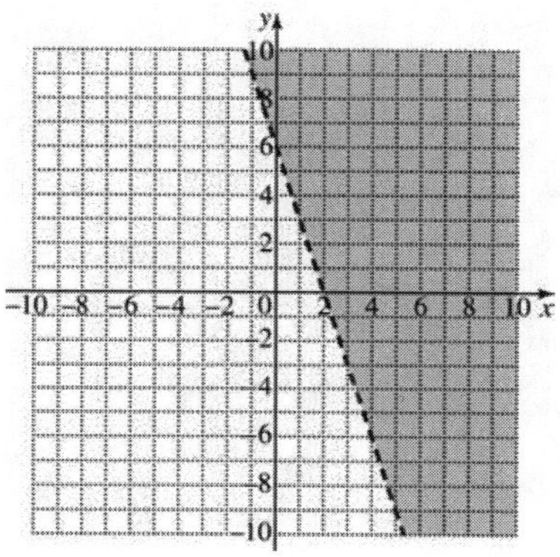

5.

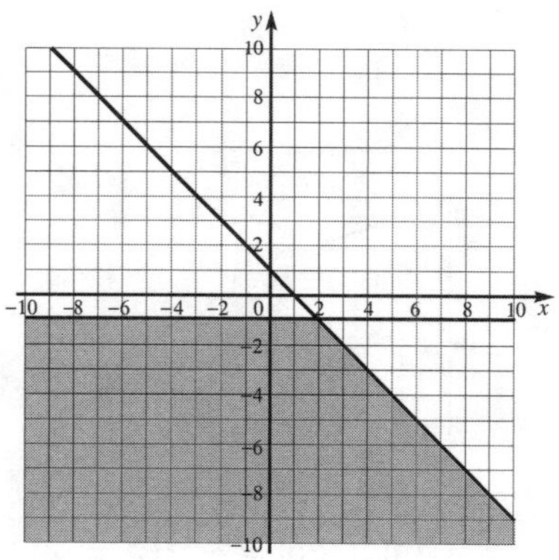

7.

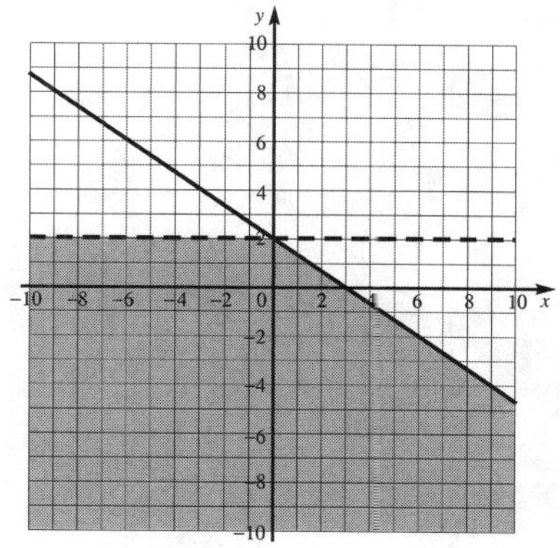

9.

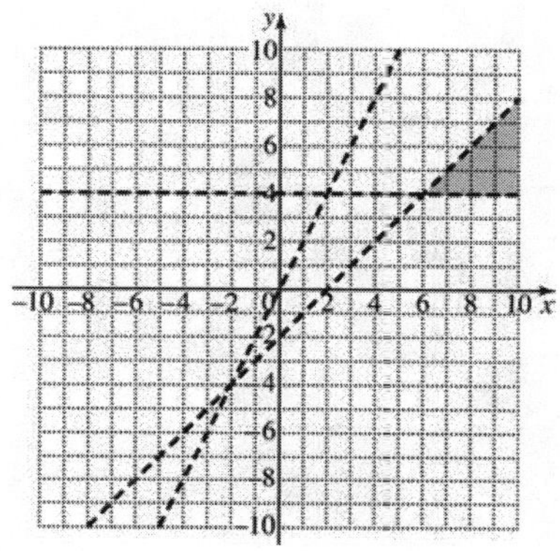

Practice Test A

1. II
2.

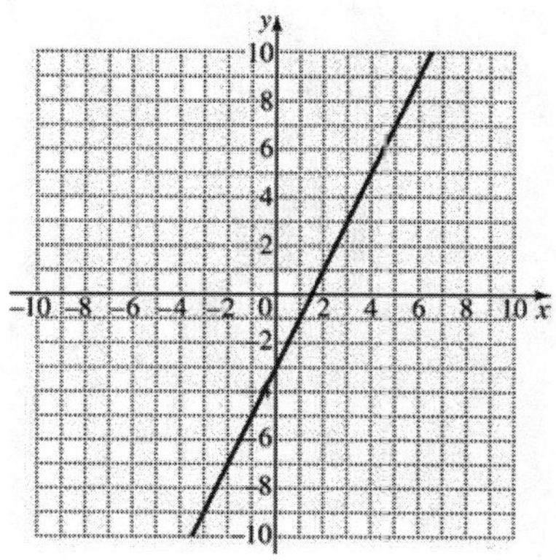

3.

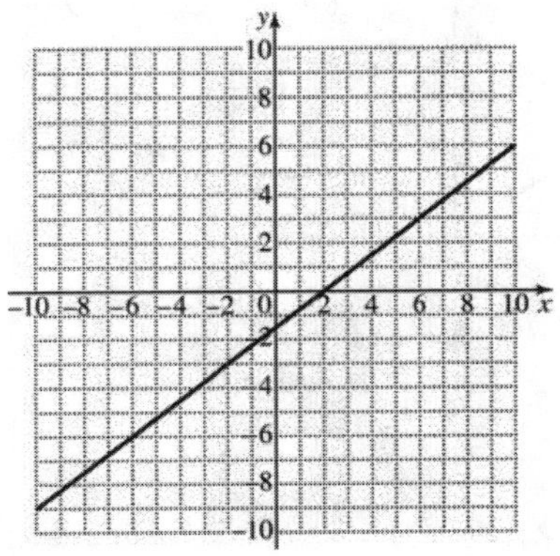

4. $-\frac{5}{2}$

5.

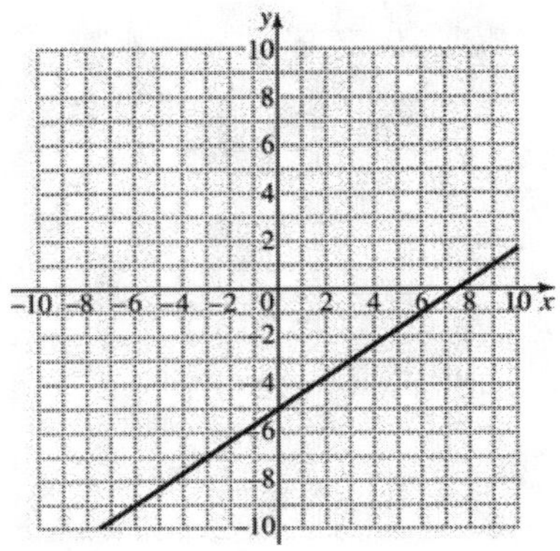

6.

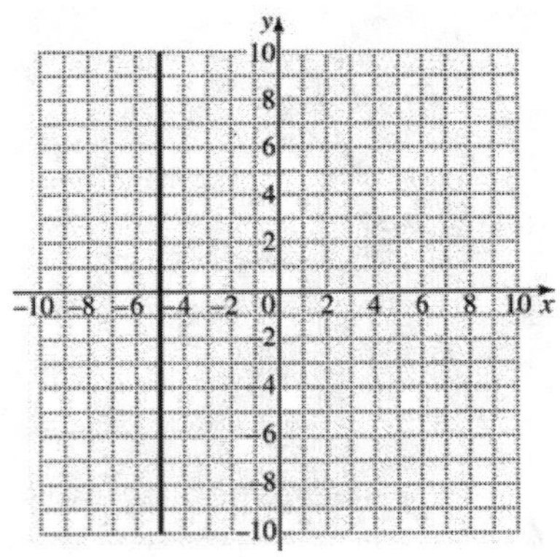

7.

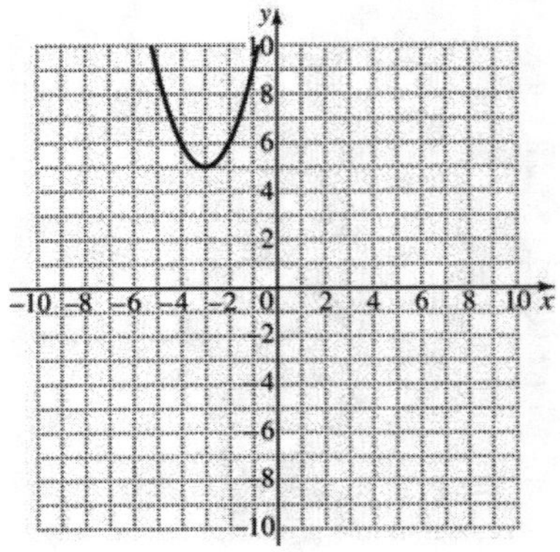

8.

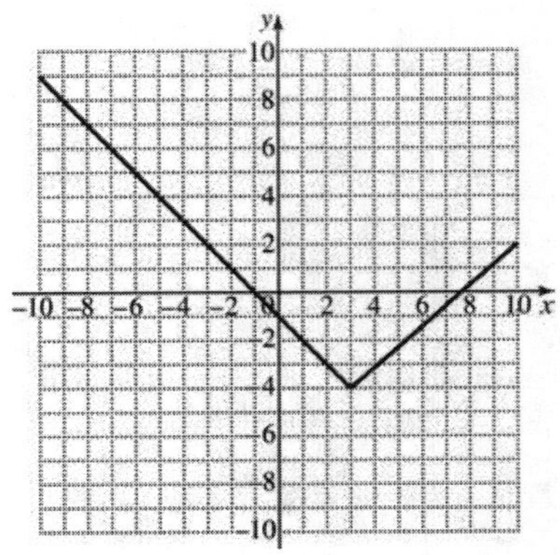

9.

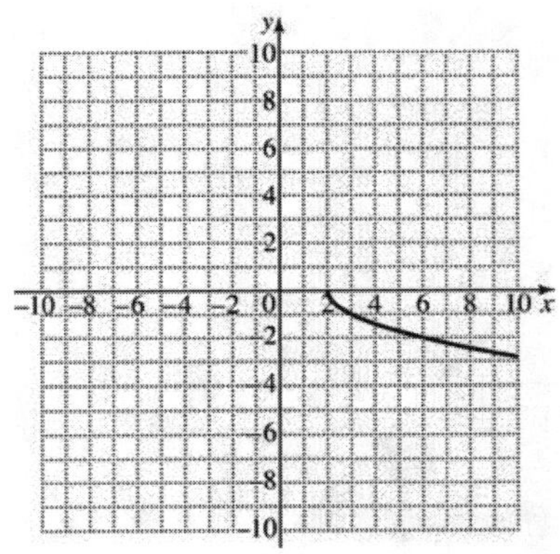

10. slope = $\frac{8}{5}$; y-intercept = $-\frac{12}{5}$
11. $3x + y = -16$
12. $x = -\frac{2}{5}$
13. $-12x + 5y = 4$ or $12x - 5y = -4$
14. $-x + 3y = -11$ or $x - 3y = 11$
15. $D = (-\infty, \infty)$; $R = [5, \infty)$; function

16.

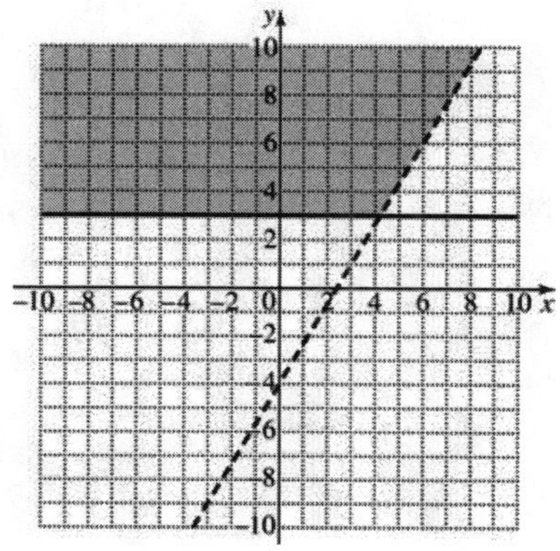

17.

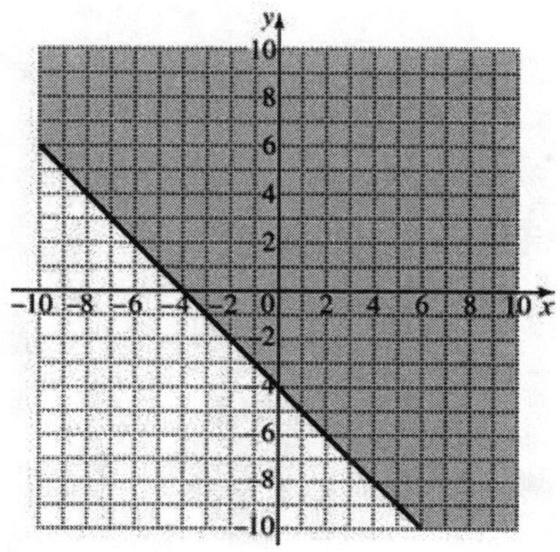

18.

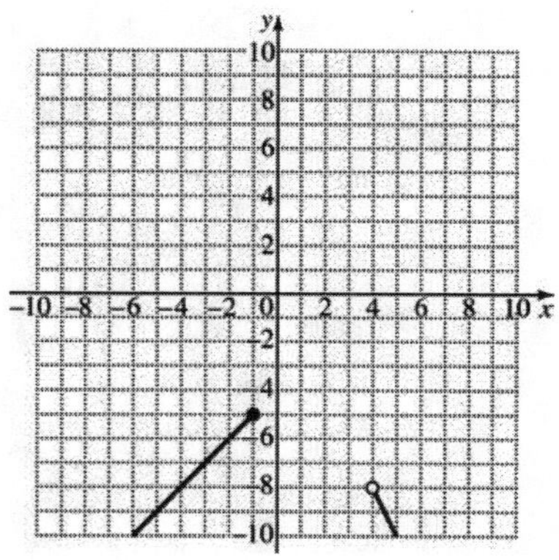

19.

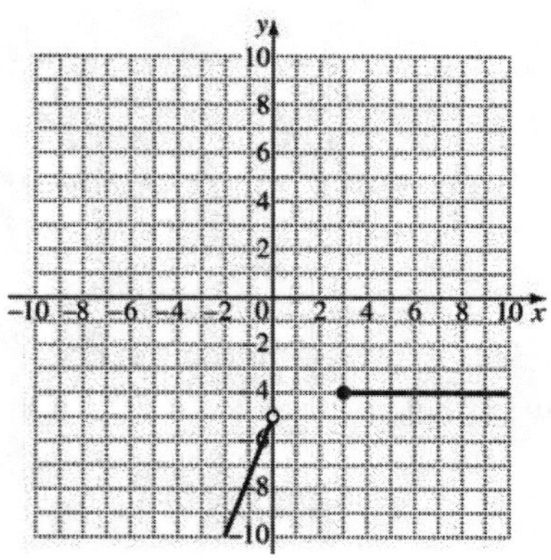

20. Parallel

Practice Test B

1. D
2. A
3. D
4. A
5. B
6. A
7. A
8. D
9. B
10. A
11. D
12. A
13. B
14. C
15. A
16. D
17. B
18. B
19. A
20. B

Chapter 4
Section 4.1

1. No
3. No
5. $(-2,8)$
7. $(15,0)$
9. $(5,2)$
11. $\{(x,y) \mid x=3y+2\}$
13. $\{(x,y) \mid 4x+3y=5\}$
15. $(0,4)$

Section 4.2

1. $(3,3,2)$
3. $\left(-\frac{1}{2},\frac{11}{2},\frac{13}{2}\right)$
5. $(3,2,2)$
7. $(1,4,1)$
9. $(3,-4,6)$
11. $(1,-2,4,-3)$

Section 4.3

1. $\{-3,-4\}$
3. 12 bracelets; 14 necklaces
5. 4 mph, and 2 mph
7. 150 units
9. After 400 rentals
11. 45°, 50°, & 85°

Section 4.4

1. Rows
3. Element
5. $(-4,2)$
7. $(2,-5)$
9. $\{(x,y) \mid 2x-5y=8\}$
11. $(1,2,-2)$
13. $(1,-1,2,3)$

Section 4.5

1.

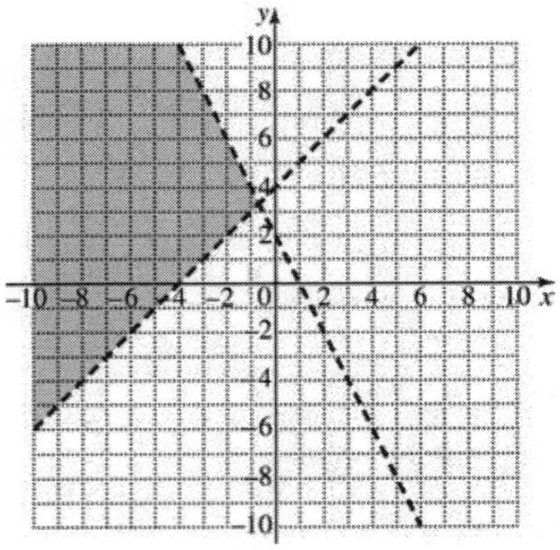

3.

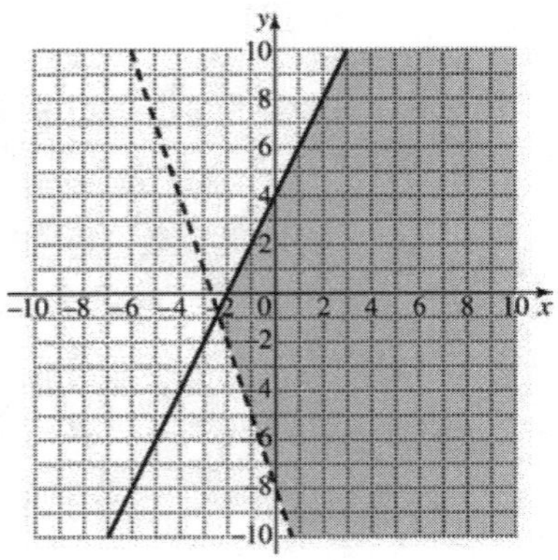

5.

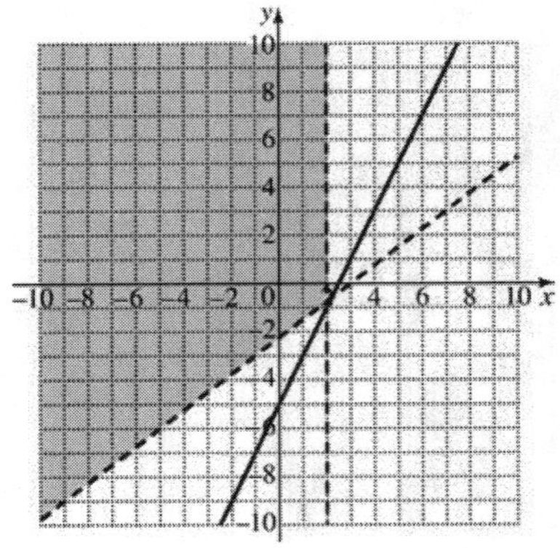

7.

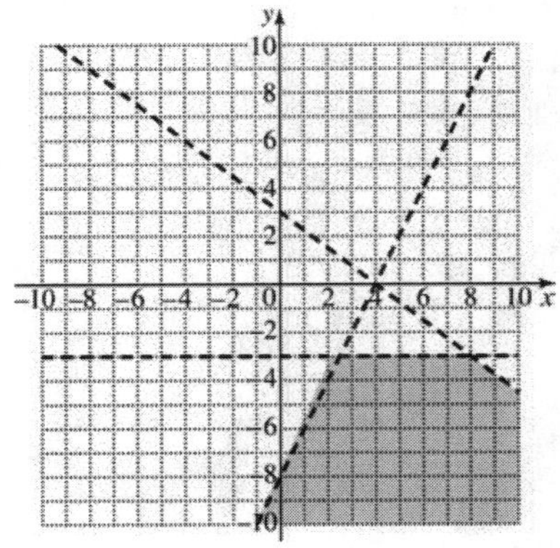

Practice Test A

1. $(6,-1)$
2. $(5,-3)$
3. $(-4,-5)$
4. $(1,-1)$
5. $(-3,4,-6)$
6. $\{(x,y) \mid 10x-12y=6\}$
7. $(-2,3)$
8. $(-1,3,4)$
9. $\varnothing$
10. $\{3,-6\}$
11. 285
12. 54 adult tickets and 28 children tickets
13. 20,000 units

14.

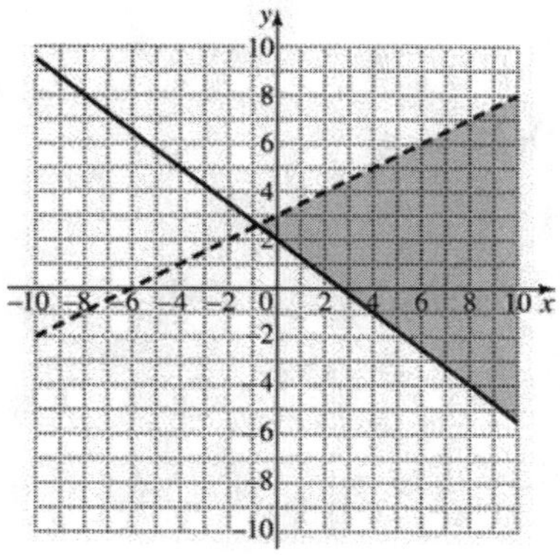

15.

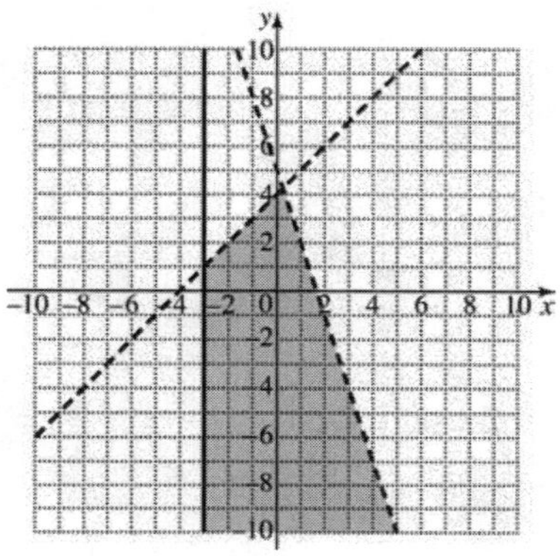

Practice Test B

1. C
2. B
3. B
4. A
5. B
6. D
7. B
8. C
9. A
10. D
11. C
12. D
13. C
14. C
15. B

Chapter 5
Section 5.1

1. $-15x^5y^7$
3. $3t^{13}$
5. -5
7. $\frac{t^4}{2}$
9. $\frac{3x^3y^2}{2}$
11. $\frac{c^3}{a^2b^4}$
13. $\frac{7}{x^3z^5}$
15. $\frac{2a^4}{z^6}$
17. 4.56×10^{-7}
19. 0.00000000000023

Section 5.2

1. $\frac{b^9c^{12}}{a^6}$
3. $\frac{64x^{12}}{z^{24}}$
5. $\frac{1}{8a^6y^3}$
7. $\frac{1}{y^{15}}$
9. $32x^{9n+3}$
11. 9×10^{11}
13. -1 and $+1$

Section 5.3

1. Trinomial
3. Binomial
5. Like
7. Coefficient
9. 12
11. 35
13. 4; none of these
15. 501
17. $6x^2-xy+16y^2$
19. $8y^2-2y-5$
21. $9t^2-5t+5$
23. $3x^2-11x+2$
25. $-13x^2+2x-1$

Section 5.4

1. $-8x^5+14x^3$
3. $x^4+x^3-x^2-5x+4$
5. $6a^2-13ab-5b^2$
7. $5x^2-x-6$
9. $16x^4-8x^2+1$
11. $25x^2-9$
13. $4s^2-12s+8$
15. y^3+2y^2-5y-6
17. $y^4-8y^3+24y^2-32y+16$
19. 2
21. $6a+3h-4$

Section 5.5

1. $5xz^3$
3. $2xy^5z^2$
5. $2x^3(3x^2-4x+1)$
7. $(x-7)(2x-3)$
9. $(3y-2)(4x-1)$
11. $(1-3b)(3+2a)$

Section 5.6

1. $(x+4)(x-2)$
3. $(x+6)(x-2)$
5. $3(x-4)(x-2)$
7. $(5x+1)^2$
9. $(2x+3)(2x+1)$
11. $(x+2)(x-7)$
13. $(x^2-5)(x^2-2)$
15. $(4x^2+3)^2$
17. -11, -4, and – 1

Section 5.7

1. $(x+4)^2$
3. $(2x-3)^2$
5. $(x+9)(x-9)$
7. $-2(4x-5y)(4x+5y)$
9. $(z-1)(z^2+z+1)$
11. $3(x+2)(x^2-2x+4)$
13. $(a^2-b^3)(a^4+a^2b^3+b^6)$
15. $(2x^n-7)(2x^n+7)$

Section 5.8

1. – 5 and 7
3. 2 and – 1
5. 4 and 1
7. $3/5$ and – 1
9. $\pm 6/5$
11. 3 and 5
13. 7 and 13 or -13 and – 7
15. $1/2$ and $-2/3$
17. $-5/2$

Practice Test A

1. $-18x^6$
2. $\frac{s^6t^6}{r^6}$
3. $\frac{8s^6u^9}{t^{12}}$
4. $\frac{1}{a^8c^5}$
5. 3.8×10^{-7}
6. 4.5×10^{9}
7. 6.6×10^{-5}
8. 1.05×10^{-6}
9. $8x^2+2x-2$
10. $4x^3+2x^2+3x+20$
11. $12x^2+x-35$
12. $225x^2+120x+16$
13. $324x^2-169$

14. $2x^3 - 3x^2 - x + 12$
15. $(x-5)(x-3)$
16. $8(x-2)(x^2+2x+4)$
17. $(c-3)(5c+3a)$
18. $4(3x+1)^2$
19. Prime
20. ± 4
21. 4 and 5
22. 5 and $-\frac{1}{2}$
23. 0, 2, and – 2
24. $\frac{3}{2}$
25. $\frac{3}{2}$ and – 1

Practice Test B

1. C
2. A
3. B
4. C
5. A
6. D
7. B
8. A
9. A
10. B
11. B
12. A
13. D
14. A
15. D
16. C
17. B
18. C
19. A
20. B
21. D
22. A
23. C
24. B
25. D

Chapter 6
Section 6.1

1. – 1
3. Domain
5. 1
7. $\{x \mid x \text{ is a real number, } x \neq -1, x \neq -2\}$
9. $\{x \mid x \text{ is a real number, } x \neq -2, x \neq 2\}$
11. $x - 4$
13. $\dfrac{x-3}{x-2}$
15. $\dfrac{(x+6)(x+4)}{x+1}$
17. $\dfrac{5(x^2+1)}{(x+2)(x-1)}$
19. -1

21a. \$100 million
21b. \$1000 million (or \$1 billion)
21c. \$900 million
21d. $\{x \mid x \text{ is a real number} > \frac{2}{3}\}$

Section 6.2

1. $\dfrac{-11}{3x}$
3. $-\dfrac{5}{x}$
5. $36x^2$
7. $x(x+2)(x-1)$
9. $(a+b)(a-b)^2$
11. $\dfrac{-3x+14}{(x+2)(x-2)}$
13. $\dfrac{3y+5}{(y+2)(y+4)(y+3)}$
15. Answers may vary.

Section 6.3

1. $\frac{3}{2}$
3. $2x + y$
5. $\dfrac{3y-4}{2y+2}$
7. $\dfrac{2-4xy}{5y+3x}$
9. $\dfrac{2b^2+3a}{b(b-a)}$

11. $\dfrac{x(45x^2-1)}{9(3x^3+1)}$

Section 6.4

1. $2x^2+4x-6$
3. $x^4-\dfrac{5x^2}{3}+2$
5. $3x+1$
7. $2x^2-9x+10-\dfrac{10}{x+5}$
9. $x^2-5x-23-\dfrac{41}{x-2}$
11. $2x^3+\dfrac{9}{2}x^2+10x+21+\dfrac{42}{x-2}$
13. 67
15. $\dfrac{5}{2}x^3-\dfrac{3}{4}x^2-\dfrac{1}{8}x+\dfrac{9}{16}-\dfrac{25}{16(2x+1)}$

Section 6.5

1. 6
3. 10
5. 1
7. $-13/7$
9. $1/2$ and $1/12$

Section 6.6

1. $f=\dfrac{-pc}{c-p}$
3. $a=\dfrac{2A-bh}{h}$
5. 1 or 7
7. 10 minutes
9. 15 mph

Section 6.7

1. $k=\dfrac{1}{7};\ y=\dfrac{1}{7}x$
3. $k=\dfrac{4}{5};\ y=\dfrac{4}{5}x$
5. $k=\dfrac{9}{8};\ y=\dfrac{9}{8}x$
7. $k=24;\ y=\dfrac{24}{x}$
9. $k=4;\ y=\dfrac{4}{x}$
11. 54 mph
13. $k=\dfrac{3}{5};\ y=\dfrac{3}{5}x^2z$
15. 22.5 tons
17. y is one-fourth its original amount.

Practice Test A

1. $\{x \mid x \text{ is a real number}, x\neq 3\}$
2. $\{x \mid x \text{ is a real number}, x\neq 3, x\neq -3\}$
3. $\dfrac{8}{x+2}$
4. $\dfrac{1}{x^2+3x+9}$
5. $\dfrac{x}{x-6}$
6. $\dfrac{x+4}{5(x+3)}$
7. $\dfrac{x-2}{x+2}$
8. $\dfrac{4(2x+5)}{x^2-25}$
9. $\dfrac{1}{(x+2)(x+5)}$
10. $\dfrac{2x(x+3)}{x^2-4}$
11. $\dfrac{-4(x^2-x-10)}{(x+2)(x+2)(x-2)}$
12. $\dfrac{(2x+5)(x-2)}{(x-1)^2}$
13. $\dfrac{8x^2-7}{30}$
14. $\dfrac{12}{3-2x^2}$

15. $\dfrac{(x-1)(2x-1)}{(x+2)^2}$

16. $4x^2+3x-6+\dfrac{5}{x}$

17. $2x+3$

18. $x^2+3x+2+\dfrac{3}{x-1}$

19. -23

20. $k=6$; $y=6x^2$

Practice Test B

1. C
2. A
3. D
4. A
5. C
6. B
7. A
8. D
9. B
10. B
11. B
12. B
13. A
14. A
15. C
16. D
17. C
18. A
19. B
20. D

Chapter 7
Section 7.1

1. index; radicand; radical sign
3. squares
5. 60
7. 0.5
9. Not a real number
11. 4.899
13. 10.198
15. 4
17. $-3x^3$
19. 3
21. Not a real number
23. y
25. $x-1$
27.

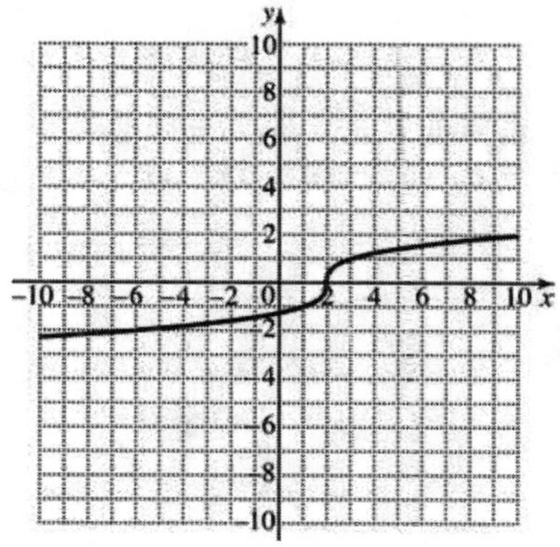

29. 20

Section 7.2

1. 5
3. $\sqrt[3]{5x}$
5. 216
7. 16
9. $\frac{1}{2}$
11. $\dfrac{4x^{3/4}}{5}$
13. $x^{3/4}$
15. $5^{7/12}$
17. $x^{5/6}+x^{1/2}-2x^{1/3}-2$
19. $x^{3/5}\left(x^{1/5}-3x^{2/5}\right)$
21. $\sqrt{y}$
23. $\sqrt{x^7y^5}$
25. $\sqrt[6]{4500}$
27. $x^{-5/12}=\dfrac{1}{x^{5/12}}$

Section 7.3

1. $\sqrt{35}$
3. $\sqrt{\dfrac{55}{6}}$
5. $\sqrt[3]{\dfrac{3}{10}}$

7. $\dfrac{9}{x\sqrt{z}}$

9. $\dfrac{z^8}{5x^6}$

11. $4x\sqrt{2}$

13. $-2x^2\sqrt[5]{y}$

15. 5; 5

17. $2\sqrt{26}$; 10.198

19. $\left(\frac{3}{2},-1\right)$

21. $\left(-\sqrt{3},\dfrac{7}{2}\sqrt{5}\right)$; $\sqrt{173}$

Section 7.4

1. $17\sqrt{2}-15\sqrt{5}$

3. $\left(y^2+3y+1\right)\sqrt{y}$

5. $\dfrac{-2\sqrt{3x}}{5}$

7. $6\sqrt{2}-12$

9. $-16-3\sqrt{5}$

11. $7-4\sqrt{3}$

13. $22\sqrt{3}$ feet

Section 7.5

1. Rationalizing the numerator

3. $\dfrac{\sqrt{5}}{\sqrt{5}}$

5. $\dfrac{\sqrt{30x}}{6x}$

7. $4\sqrt[5]{4x}$

9. $\dfrac{\sqrt{10}+4\sqrt{5}+\sqrt{6}+4\sqrt{3}}{14}$

11. $\dfrac{7\left(3+\sqrt{x}\right)}{9-x}$

13. $\dfrac{x}{5\sqrt{3x}}$

15. $\dfrac{1-x}{3\left(1-\sqrt{x}\right)}$

17. $\dfrac{2-16x^2}{4x^2+2+5x\sqrt{2}}$

Section 7.6

1. 35

3. $\varnothing$

5. – 12

7. 0

9. 1

11. 202 miles

13. 1 and 0

Section 7.7

1. Imaginary Unit

3. Pure imaginary

5. Real

7. $6i$

9. $i\sqrt{5}$

11. $7-i$

13. $4+9i$

15. 60

17. $-15-23i$

19. 20

21. $\dfrac{8}{5}+\dfrac{4}{5}i$

23. $4+i$

25. – 1

27. $64i$

29. $\dfrac{1}{2}-\dfrac{3\sqrt{2}}{7}i$

Practice Test A

1. $8\sqrt{5}$

2. $2\sqrt[5]{x^4}$

3. $\frac{3}{2}$

4. $x^{7/6}-x^{3/4}$

5. a^6b^9

6. $\dfrac{7x^{1/3}}{y^{2/5}}$

7. $3a\sqrt{3i}$

8. $8i$
9. $\frac{4\sqrt{5}}{5}$
10. $\frac{x+8\sqrt{x}+15}{9-x}$
11. $\frac{\sqrt[3]{x^2}}{x}$
12. $\frac{21}{37}+\frac{22}{37}i$
13. $30 - 4\sqrt{2}$
14. $\sqrt{15}-\sqrt{10}$
15. $7-2x\sqrt{7}+x^2$
16. $13x\sqrt{2}+2\sqrt{2x}$
17. 13. 115
18. 4.611
19. 22
20. 6 and – 1
21. 7
22. $27+11i$
23. $1-3i$
24. $\sqrt{146}$
25. $\left(\frac{5}{2},2\right)$

Practice Test B

1. A
2. B
3. D
4. B
5. C
6. A
7. A
8. B
9. A
10. D
11. A
12. C
13. A
14. B
15. A
16. D
17. A
18. C
19. A
20. C
21. A
22. D
23. B
24. C
25. D

Chapter 8
Section 8.1

1. Quadratic Equation
3. Completing the Square
5. 6 and – 6
7. $\pm\sqrt{5}$
9. $3\pm 3i$
11. $-3\pm\sqrt{7}$
13. $\frac{-3\pm i\sqrt{11}}{2}$
15. $-3\pm\sqrt{3}i$
17. $-\frac{3}{2}\pm\frac{\sqrt{13}i}{2}$
19. 40%
21. $6\sqrt{2}$

Section 8.2

1. – 15 and – 1
3. $\frac{-3\pm\sqrt{57}}{4}$
5. $-\frac{1}{4}\pm\frac{\sqrt{39}}{4}i$
7. Two real solutions
9. Two complex solutions
11. 14 ft
13. $\frac{-\sqrt{5}\pm\sqrt{17}}{2}$

Section 8.3

1. 5 and 1
3. 3
5. 1 and $-\frac{1}{2}$
7. $-\frac{20}{3}$
9. $\pm 1, \pm 2i$
11. $7-2\sqrt{6}$
13. 5 mph, then 4mph

15. Joe will take 15.1 hours, and Sarah will take 17.1 hours

17. -1 and $-\frac{1}{5}$

Section 8.4

1. $[-4,3]$
3. $(-6,2)$
5. $(-\infty,-5)\cup(-1,1)\cup(5,\infty)$
7. $(-\infty,-3)\cup\left(-\frac{8}{3},\infty\right)$
9. $(-7,2)\cup\left(\frac{14}{5},\infty\right)$
11. Answers will vary.

Section 8.5

1. axis of symmetry at x = 0   Vertex $(0,2)$

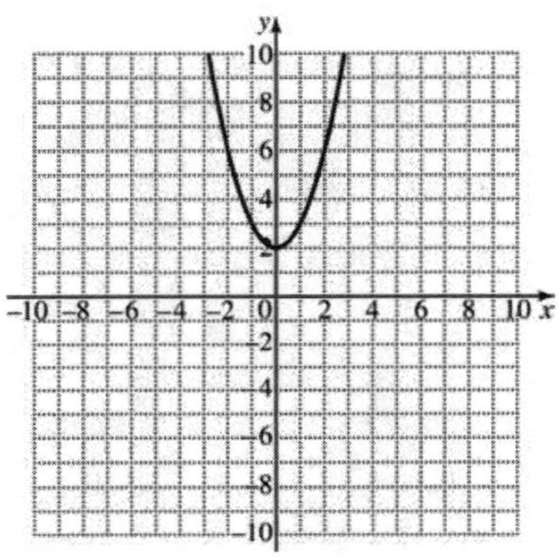

3. axis of symmetry at x = 0   Vertex $(0,6)$

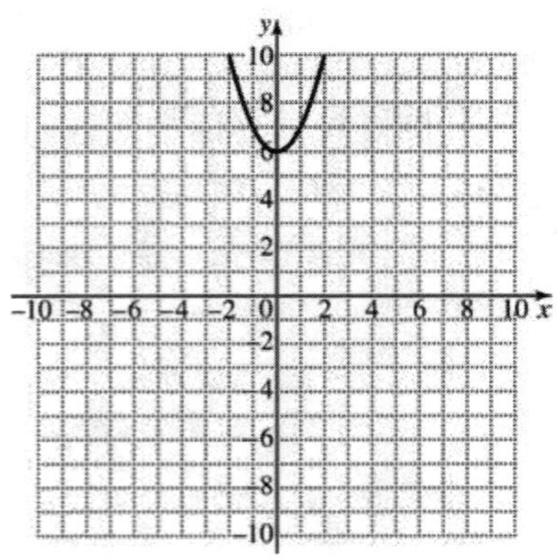

5. axis of symmetry at $x=4$ Vertex $(4,0)$

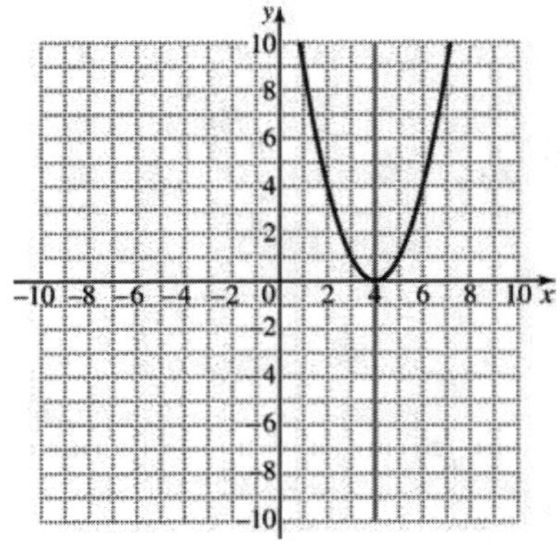

7. axis of symmetry at $x=2$ Vertex $(2,5)$

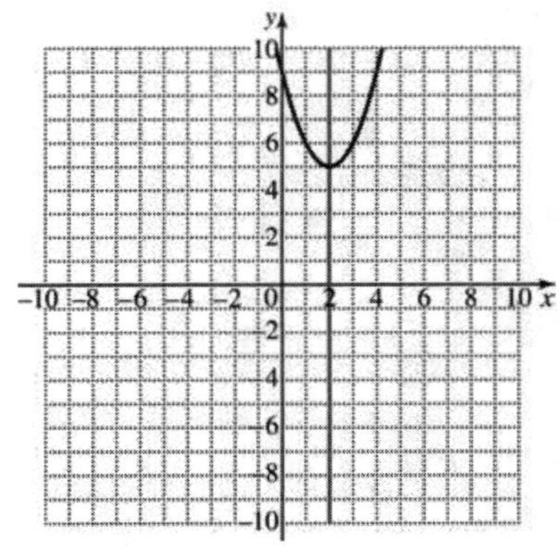

9. axis of symmetry at $x=3$ Vertex $(3,-2)$

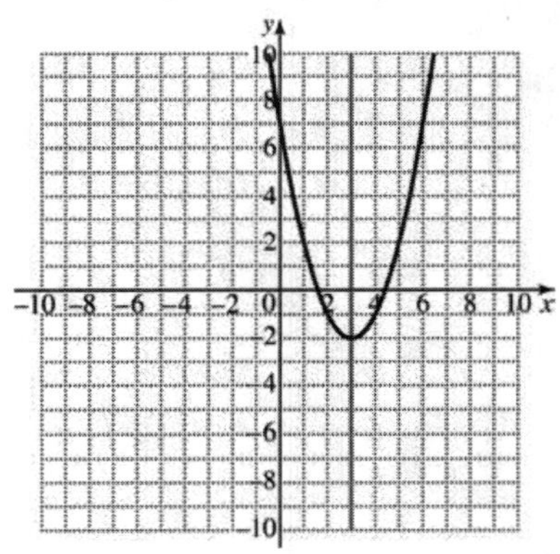

11. axis of symmetry at $x=0$ Vertex $(0,0)$

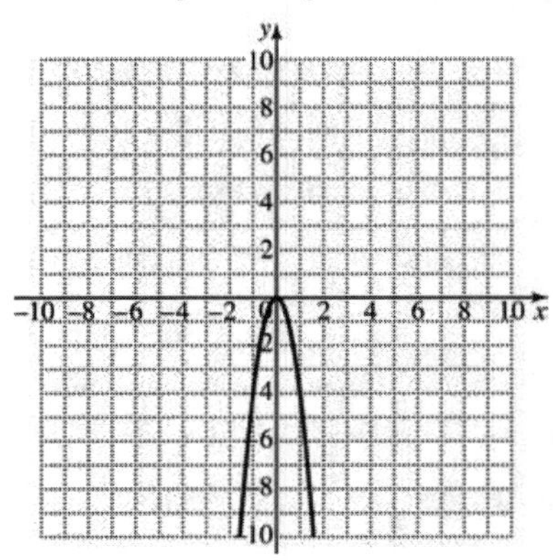

13. axis of symmetry at $x = 0$ Vertex $(0,0)$

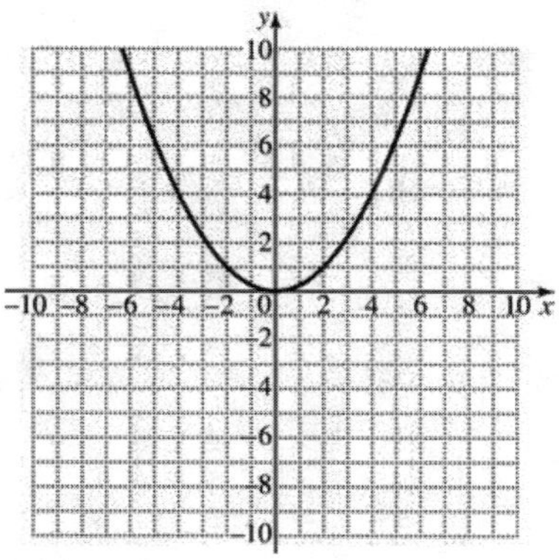

15. axis of symmetry at $x = 1$ Vertex $(1,2)$

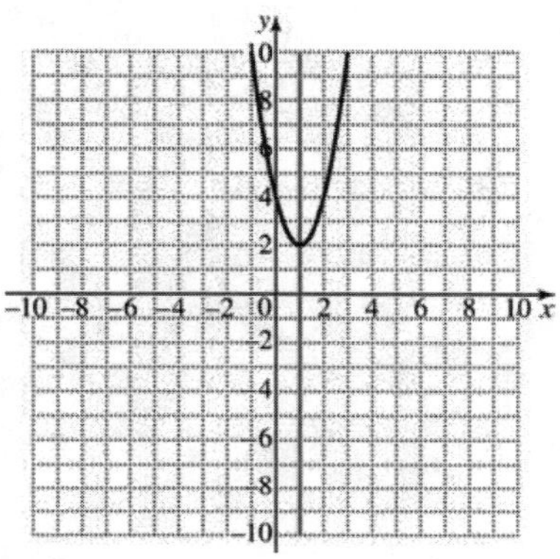

17. $f(x) = -2(x-4)^2 - 6$

Section 8.6

1. $f(x) = (x-1)^2 - 16$; $(1,-16)$
3. $f(x) = \left(x + \frac{5}{2}\right)^2 + \frac{47}{4}$; $\left(-\frac{5}{2}, \frac{47}{4}\right)$
5. $f(x) = -3(x-2)^2 + 20$; $(2,20)$
7. $-\dfrac{b}{2a}$
9. $(-2,3)$
11. $(1,-16)$
13. 144 ft
15. 1.8; minimum

Practice Test A

1. $\dfrac{3 \pm \sqrt{29}}{2}$
2. $\frac{2}{7}$ and $-\frac{8}{5}$
3. $-3 \pm \sqrt{6}$
4. 13 and 1
5. ± 3 and $\pm i$
6. 9 and 11
7. 3, - 3, and - 2
8. $3 \pm 2\sqrt{5}$
9. $6 \pm \sqrt{17}$
10. $\dfrac{6 \pm \sqrt{21}}{2}$
11. $\left(-\infty, -\frac{5}{3}\right] \cup [8, \infty)$
12. $\left(-7, \frac{9}{2}\right)$
13. $\left(-\infty, -\frac{3}{2}\right) \cup (1, \infty)$
14. Vertex $(-3,6)$

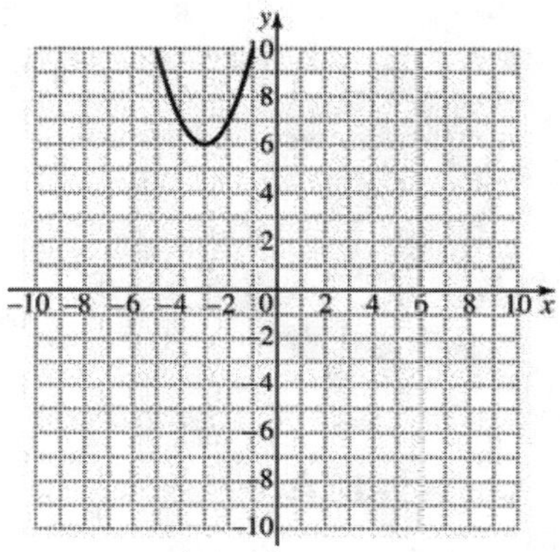

15. Vertex $(0,-4)$

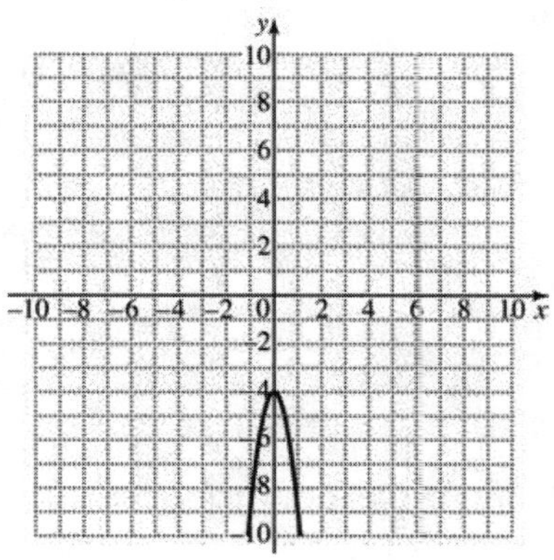

16. Vertex $(-2,5)$

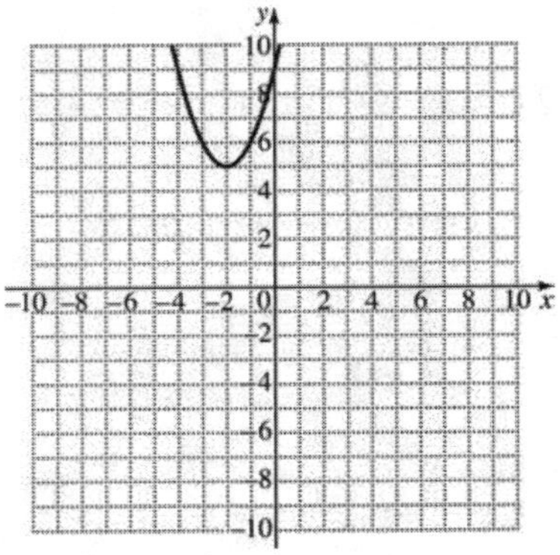

17. Vertex $(4,7)$

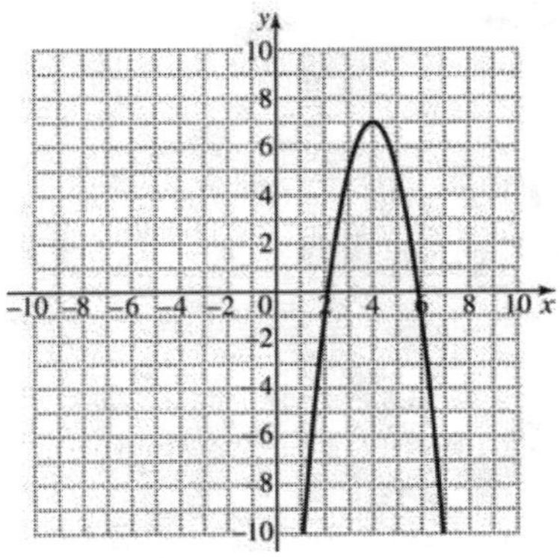

18. 13 and 21
19. height is 4 ft, base is 16 ft.
20. 4.5 seconds

Practice Test B

1. A
2. D
3. B
4. A
5. A
6. C
7. D
8. B
9. C
10. A
11. A
12. C
13. A
14. C
15. B
16. D
17. A
18. B
19. C
20. A

Chapter 9
Section 9.1

1 a. $3x-2$, b. $-x-8$

c. $2x^2-7x-15$ d. $\dfrac{x-5}{2x+3}$

3 a. $\sqrt{x+5}+4x^2$ b. $\sqrt{x+5}-4x^2$

c. $4x^2\sqrt{x+5}$ d. $\dfrac{\sqrt{x+5}}{4x^2}$

5. 14
7. 23
9. $\sqrt{x+3}-7;\ \sqrt{x-4}$
11. $(2x+5)^{2/3}-9;\ \sqrt[3]{2x^2-13}$
13. 6
15. 18
17. -6

Section 9.2

1. $y=x$
3. Horizontal
5. x
7. No
9. No
11. $4x-3$
13. x^3+5
15. $f^{-1}(x)=\dfrac{x-1}{5}$

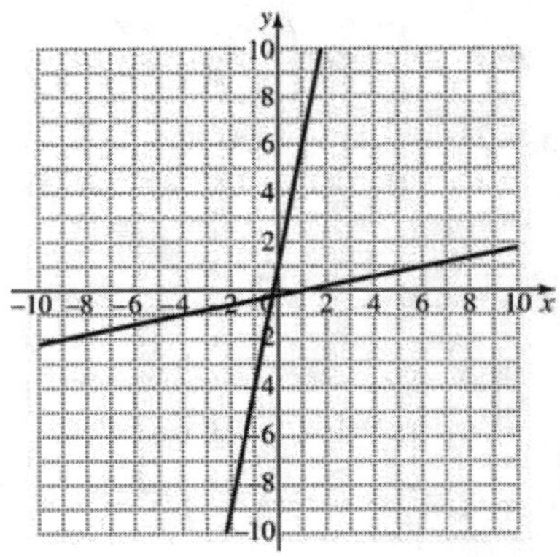

17a. $(-5,25)$ b. $(25,-5)$

Section 9.3

1.

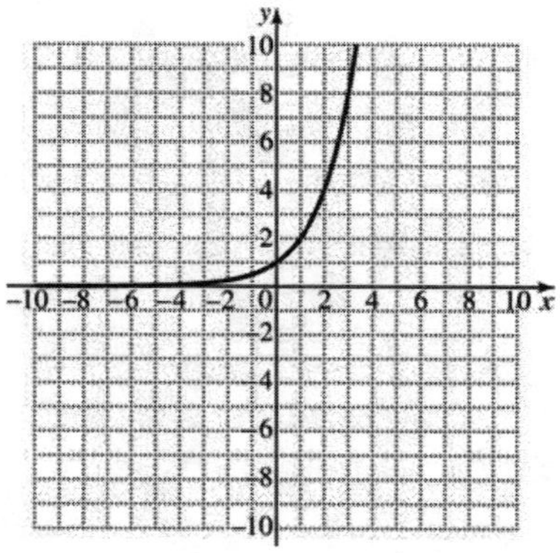

3.

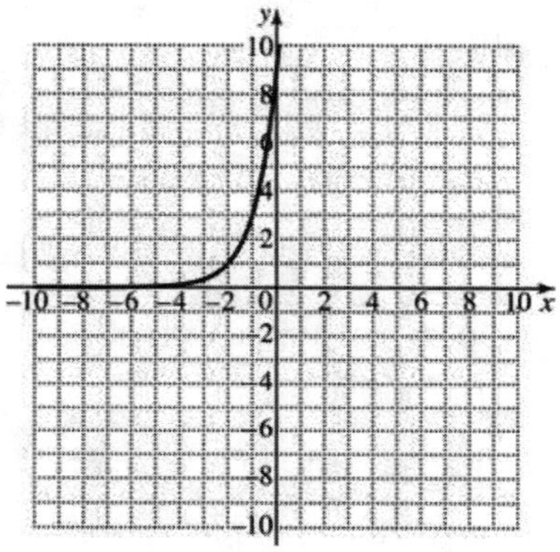

5. -3 7. $-\frac{1}{3}$

9. $182,250

11.

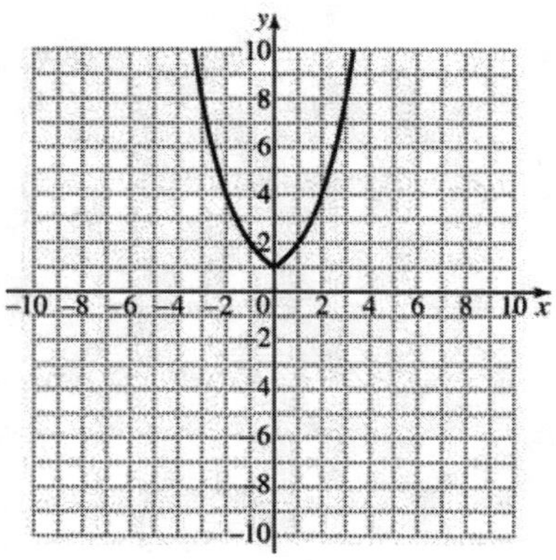

Section 9.4

1. $4^2 = 16$

3. $7^9 = x$

5. $\log_5 \frac{1}{125} = -3$

7. $\log_3 81 = 4$

9. 5

11. -2

13. -3

15. 8

17. 3

19.

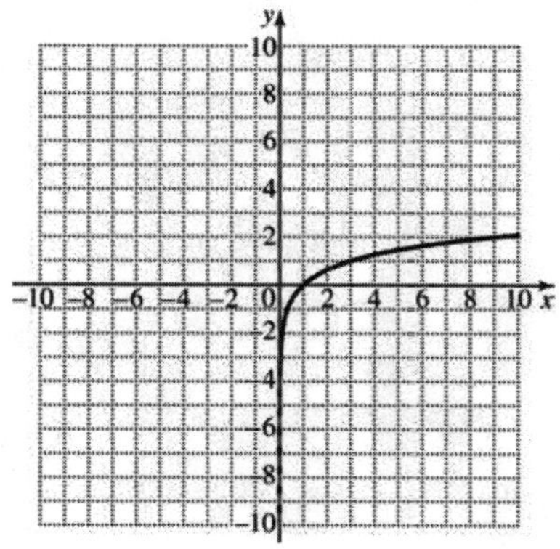

21.

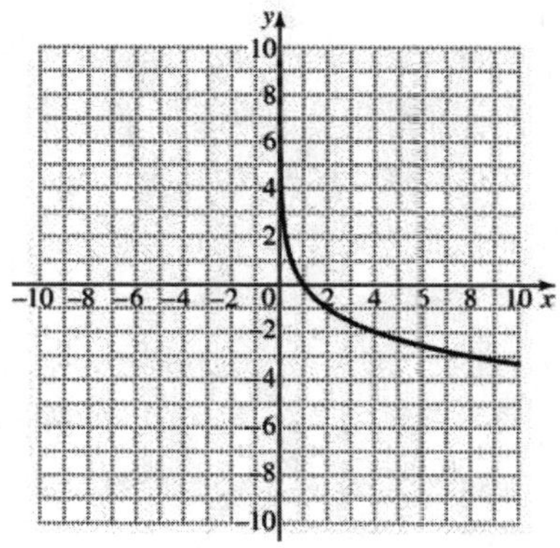

Section 9.5

1. $\log_3 5x$

3. $\log_7 \left(x^5 - 2x^4\right)$

5. $\log_3 \frac{x}{6}$

7. $\log_2 \frac{x^2 - 6}{x + 5}$

9. $6 \log_4 x$

11. $\frac{1}{2} \log_5 y$

13. $\log_4 4 = 1$

15. $\log_8 4x$

17. $\log_3 5 + 2 \log_3 x$

19. $\frac{1}{2}\log_7 x - 2\log_7(x+4)$

21. $b = 3$

Section 9.6

1. 0.4082
3. 3
5. – 4
7. $\frac{10^{1.1}}{2}$; 6.2946
9. 1.6094
11. 0.02
13. – 5
15. 1.5850
17. – 2.1053

Section 9.7

1. $\frac{\ln 9}{\ln 5}$; 1.3652
3. $\frac{\ln 24}{3 \cdot \ln 6}$; 0.5912
5. 4
7. $-\frac{7}{8}$
9. $\frac{1}{4}$
11. 14 years
13. $x \le 5$

Practice Test A

1. $x - 4$
2. $-6x^2 + 8x$
3. $-6x - 4$
4. -16
5. $2x + 14$
6. $\sqrt[3]{x+4}$
7. $\frac{7x-8}{5}$
8. $\log_6 \frac{3x^2}{x+2}$
9. $\log_3 \frac{5\sqrt{x}}{8}$
10. 1.5440
11. 1024
12. $\frac{1}{5}$
13. 4 and – 2
14. $\frac{23}{8}$
15. 5.1828
16.

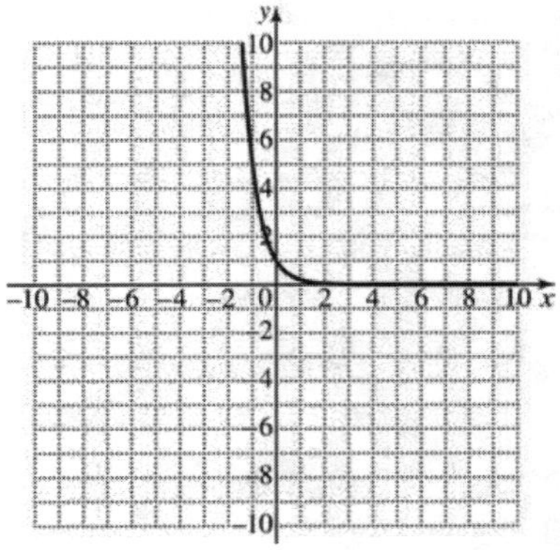

17.

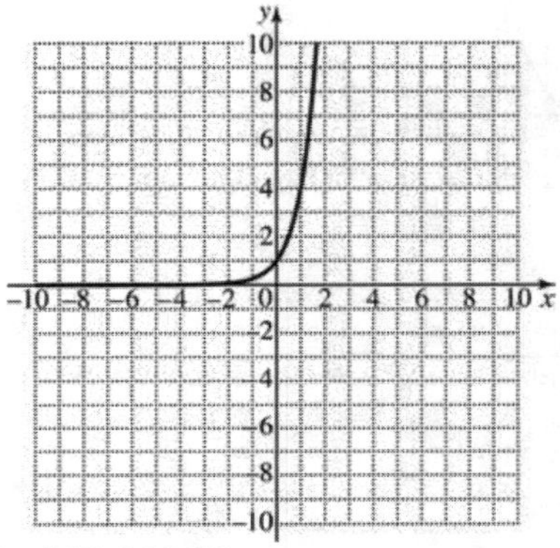

18. \$10,810.59
19. \$10,023.24
20. 0.375 grams

Practice Test B

1. B
2. C
3. A
4. A
5. D
6. C
7. B
8. C
9. A
10. B
11. C
12. A
13. D
14. B
15. A

16. C
17. B
18. D
19. C
20. A

Chapter 10
Section 10.1

1. Radius
3. Conic sections
5. Circle; center
7. V: $(-1,-4)$

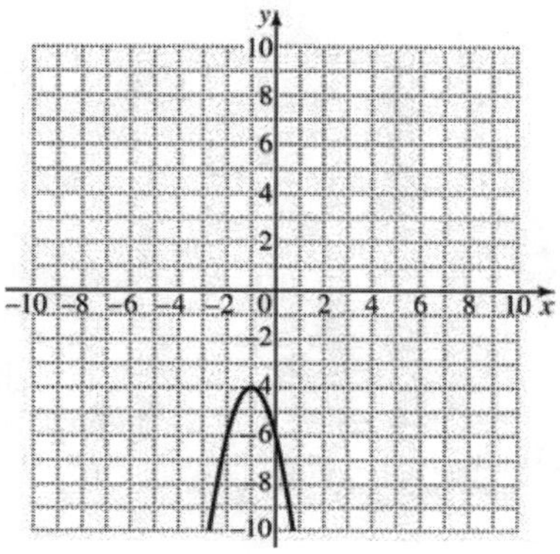

9. V: $(3,-6)$

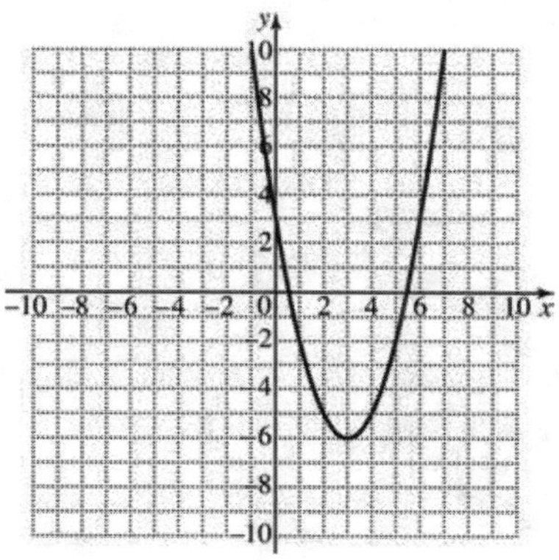

11. C: $(0,3)$; r = 2

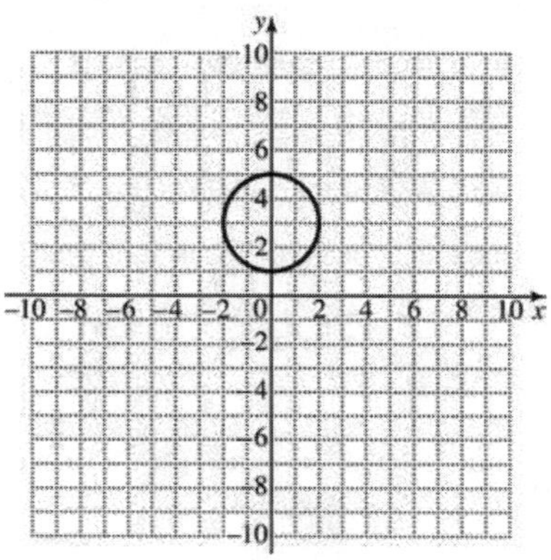

13. C: $(-2,-1)$; r = 4

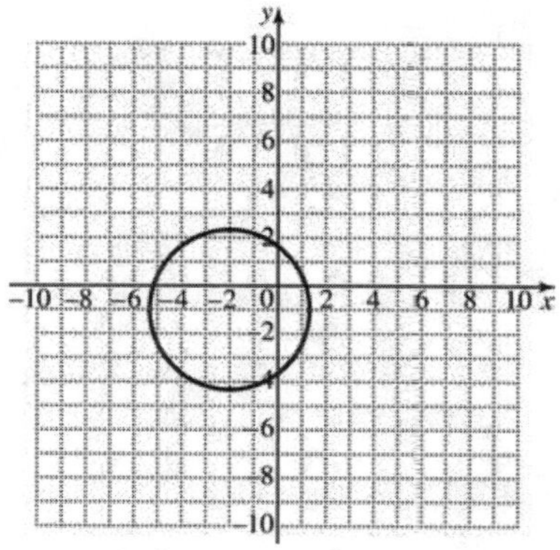

15. $(x+1)^2+(y+6)^2=49$
17. $C:(0,0) \quad r=4$
19. $C:(3,5) \quad r=\sqrt{38}$
21. If $a<0$ and x-coordinate in vertex is negative and if $a>0$ and the x-coordinate in the vertex is positive.

Section 10.2

1. Foci
3. Hyperbola ; $(0,a)$ and $(0,-a)$
5.

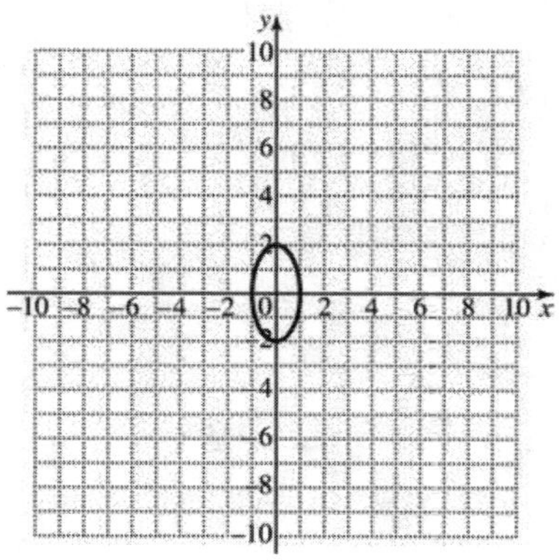

7.

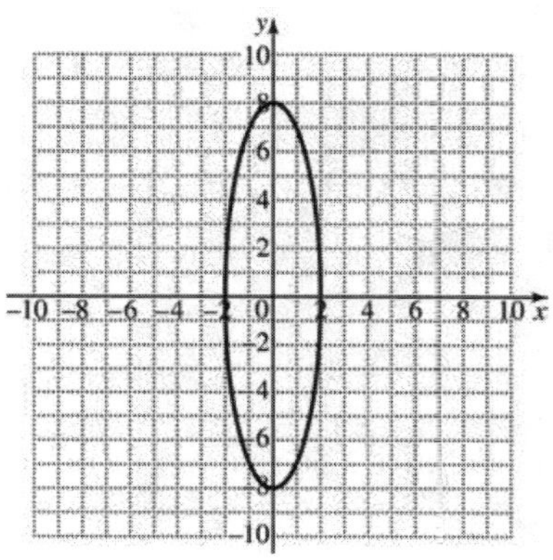

9.

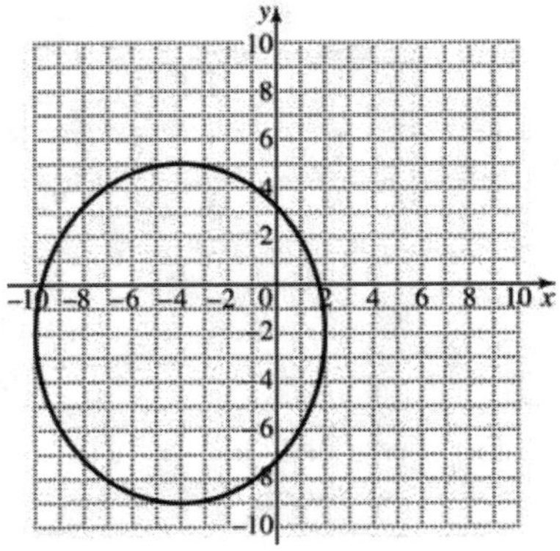

11.

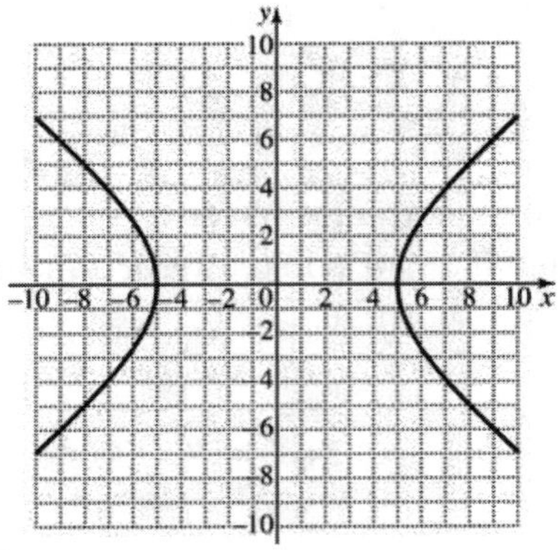

13.

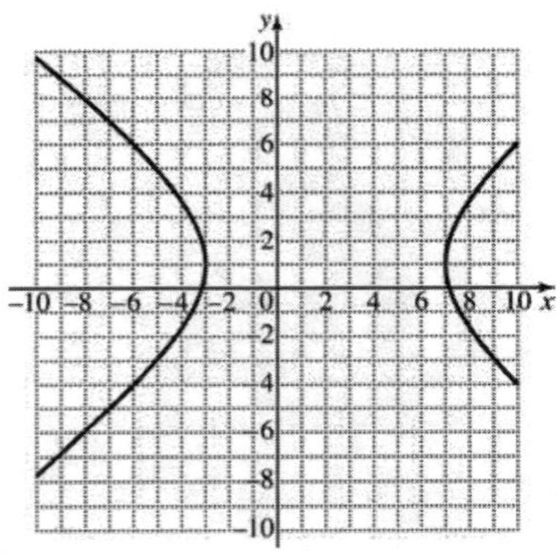

15.

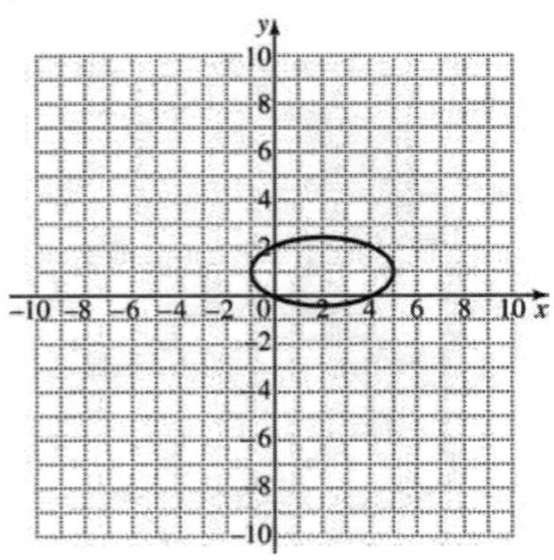

Section 10.3

1. $(0,-2)(4,0)$
3. $\varnothing$
5. $\left(\frac{12}{5},-\frac{8}{5}\right)(4,0)$
7. $\varnothing$
9. $(0,-2)\left(\frac{-4\sqrt{5}}{3},\frac{4}{3}\right)\left(\frac{4\sqrt{5}}{3},\frac{4}{3}\right)$
11. $-\frac{18}{5}$ and $\frac{24}{5}$ or 6 and 0

Section 10.4

1.

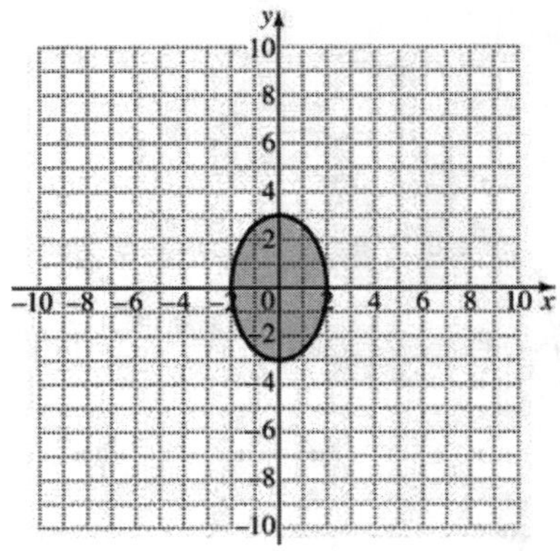

3.

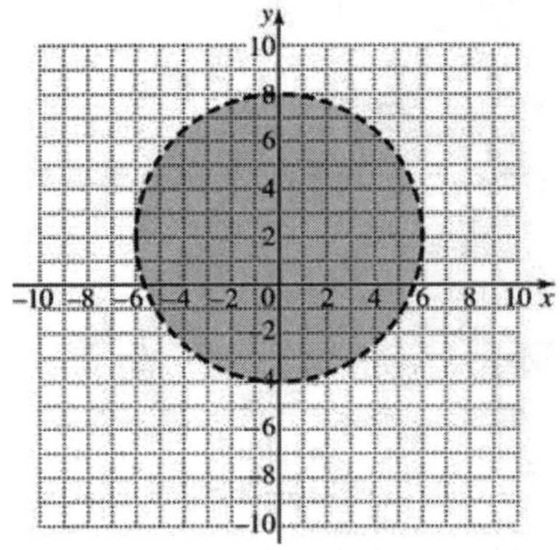

5.

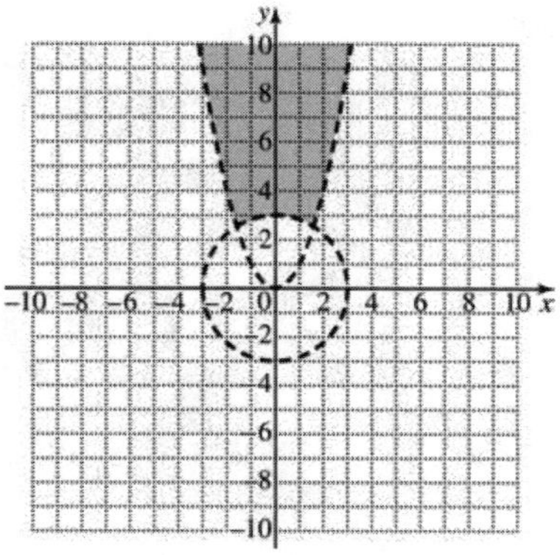

7.

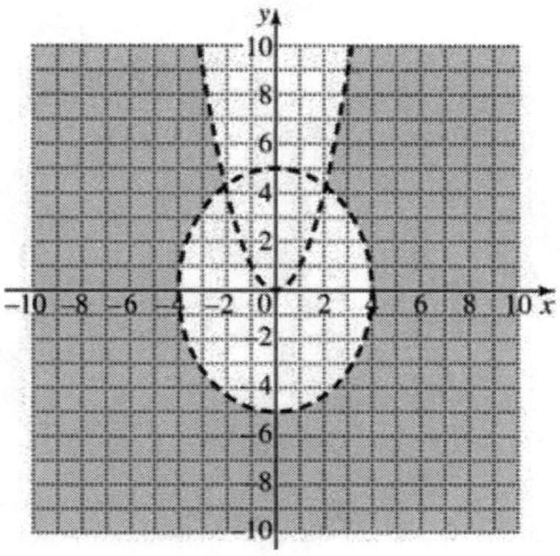

9.

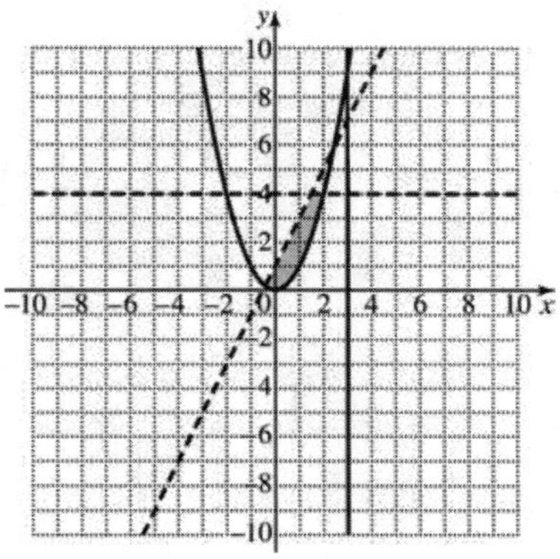

Practice Test A

1.

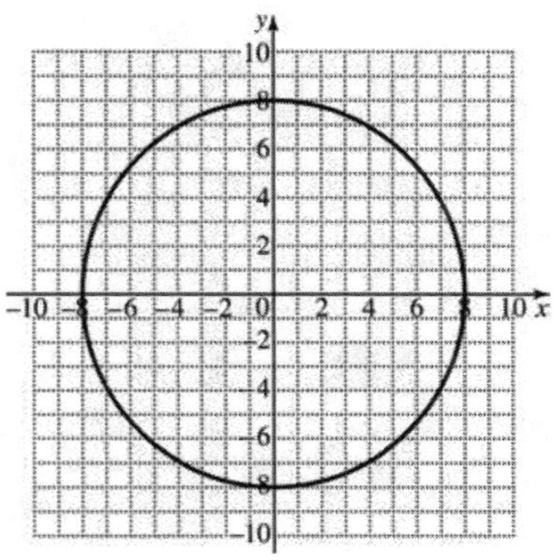

2.

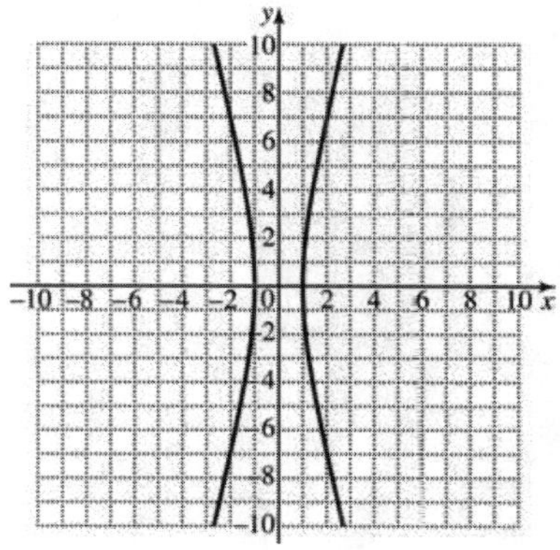

3.

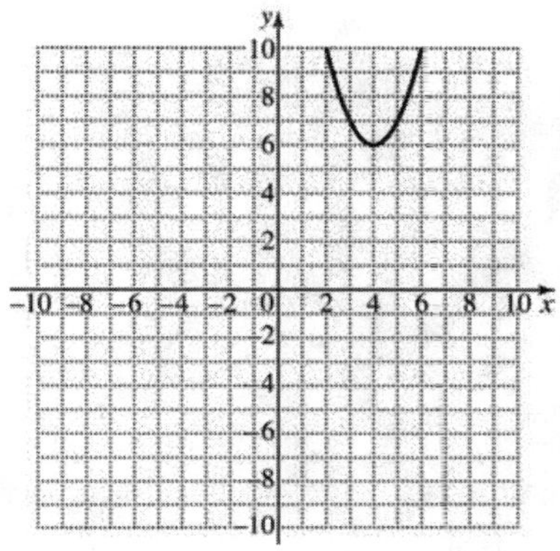

4.

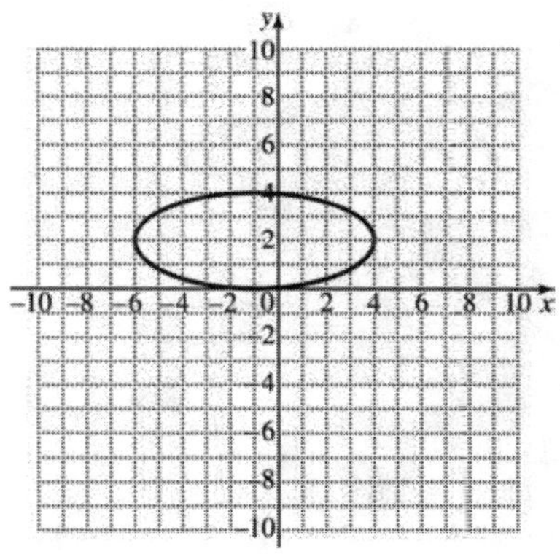

5.

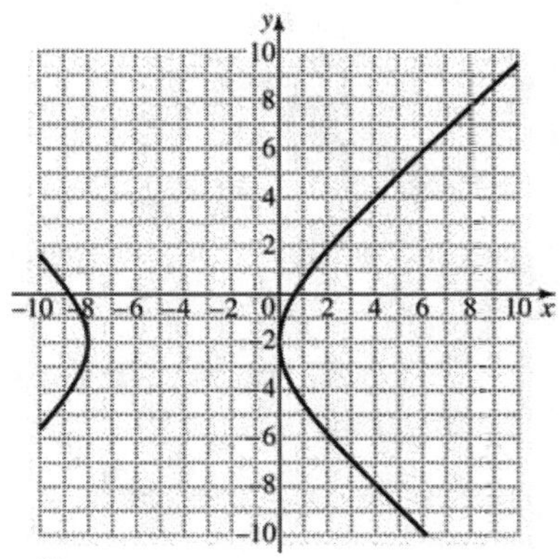

6. $\varnothing$

7. $(0,-2)(2,0)$

8. ∅

9.

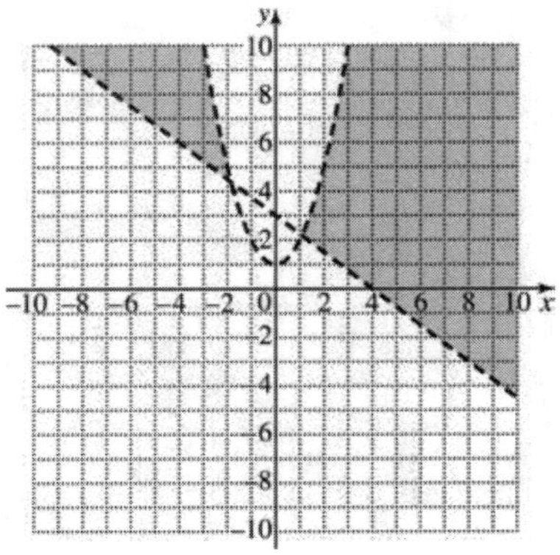

10.

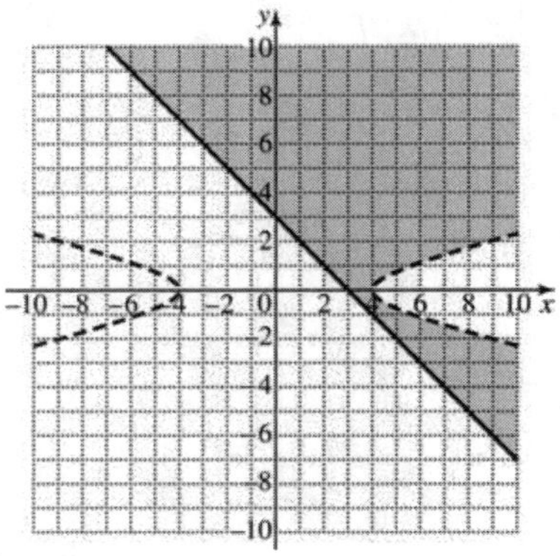

11.

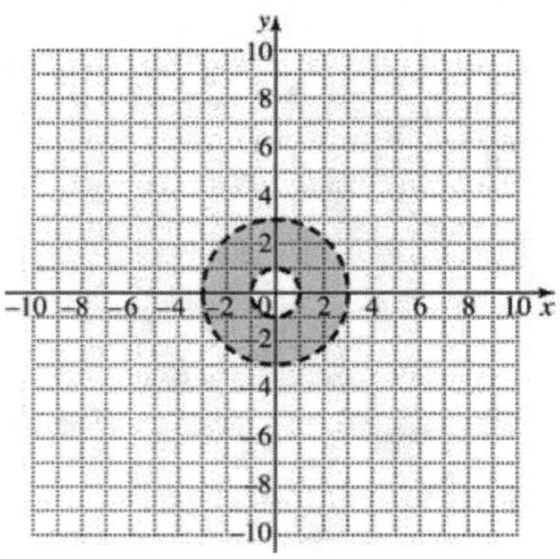

12. 11.1 feet

Practice Test B

1. A
2. B
3. C
4. A
5. D
6. B
7. C
8. D
9. A
10. D
11. B
12. A

Chapter 11
Section 11.1

1. Infinite
3. Finite
5. $1, -3, 9, -27, 81$
7. $\frac{1}{4}, \frac{1}{5}, \frac{1}{6}, \frac{1}{7} \frac{1}{8}$
9. -18
11. $\frac{-1}{400}$
13. 8388608
15. $a_n = -2^n$
17. $a_n = \frac{3}{n+4}$
19. $120, $160, $200, $240, $280
21. $1, 1.0125, 1.0167, 1.0189, 1.0202$

Section 11.2

1. Arithmetic; Difference
3. First
5. 25,32,39,46,53
7. $10, 26, 42, 58, 74$
9. 34
11. – 28
13. $48, 24, 12, 6, 3$
15. $-1, -4, -16, -64, -256$
17. – 1458
19. $-\frac{1}{32}$
21. $32.3, -71.06, 156.332, -343.9304$

Section 11.3

1. 60
3. 250
5. 496
7. $\sum_{i=2}^{6} 2i + 1$
9. – 13
11. 28

13. $\sum_{i=1}^{5} i$

Section 11.4

1. 63
3. 42
5. 484
7. $85/64$
9. 64
11. $32/9$
13. $4/9$

Section 11.5

1. $m^3+3m^2n+3mn^2+n^3$
3. $x^8+8x^7y+28x^6y^2+56x^5y^3$ $+70x^4y^4+56x^3y^5+28x^2y^6+8xy^7+y^8$
5. 255024
7. $x^5-5x^4y+10x^3y^2-10x^2y^3+5xy^4-y^5$
9. $625a^4-1000a^3b+600a^2b^2-160ab^3+16b^4$
11. $700000x^4z^4$
13. $189m^5n^2$
15. $840x^3$

Practice Test A

1. $\frac{3}{5},\frac{5}{6},1,\frac{9}{8},\frac{11}{9},\frac{13}{10}$
2. $\frac{5}{6561}$
3. $a_n=7\left(-\frac{1}{7}\right)^{n-1}$
4. $a_n=24+(n-1)(-14)$
5. 161
6. $-21/8$
7. $864/5$
8. -40
9. 88
10. 3276
11. $16x^4-32x^3y+24x^2y^2-8xy^3+y^4$
12. $a^5+15a^4b+90a^3b^2+270a^2b^3+405ab^4+243b^5$

Practice Test B

1. D
2. A
3. B
4. A
5. C
6. B
7. D
8. B
9. C
10. A
11. D
12. B